广东省文学艺术界联合会
广东省文艺研究所
广州十三行博物馆
粤海风杂志社
主编

艺述大湾区

Stories of Arts in the Greater Bay Area

广州十三行故事

The Thirteen Hongs in Canton

（2021）

广州新华出版发行集团
广州出版社

图书在版编目（CIP）数据

广州十三行故事：汉、英 / 广东省文学艺术界联合会等主编．—广州：广州出版社，2020.9

（艺述大湾区）

ISBN 978-7-5462-3107-5

Ⅰ．①广… Ⅱ．①广… Ⅲ．①历书—中国—2021 ②十三行—史料—汉、英 Ⅳ．① P195.2 ② F752.949

中国版本图书馆 CIP 数据核字 (2020) 第 177879 号

艺述大湾区：广州十三行故事

Yishu Dawan Qu Guangzhou Shisanhang Gushi

出版发行　广州出版社

（地址：广州市天河区天润路 87 号 9、10 楼

邮政编码：510635 网址：www.gzcbs.com.cn）

责任编辑　李素娟

责任校对　张　岚 李　珊 李少芳 黄焕姗

装帧设计　榕汇设计

印　　刷　广州一龙印刷有限公司

（地址：广州市增城区新塘镇荔新九路 43 号千亿产业园

邮政编码：511340）

规　　格　889mm × 1194mm 1/32

印　　张　24

字　　数　360 千

版　　次　2020 年 9 月第 1 版

印　　次　2020 年 9 月第 1 次

书　　号　978-7-5462-3107-5

定　　价　128.00 元

目录

Orders to record

序言

Preface

广州：岭南华府，千年商都

Canton: The Metropolis of Lingnan, a Commercial Hub of Two Thousand Years

2021

农历辛丑年（牛年）

1 January

一	二	三	四	五	六	日
				1 元旦	2 十九	3 二十
4 廿一	5 小寒	6 廿三	7 廿四	8 廿五	9 廿六	10 廿七
11 廿八	12 廿九	13 腊月	14 初二	15 初三	16 初四	17 初五
18 初六	19 初七	20 大寒	21 初九	22 初十	23 十一	24 十二
25 十三	26 十四	27 十五	28 十六	29 十七	30 十八	31 十九

2 February

一
1 二十
8 廿七
15 初四
22 十一

5 May

一	二	三	四	五	六	日
					1 劳动节	2 廿一
3 廿二	4 五四青年节	5 立夏	6 廿五	7 廿六	8 廿七	9 廿八
10 廿九	11 三十	12 四月	13 初二	14 初三	15 初四	16 初五
17 初六	18 初七	19 初八	20 初九	21 小满	22 十一	23 十二
24 十三	25 十四	26 十五	27 十六	28 十七	29 十八	30 十九
31 二十						

6 June

一
7 廿七
14 端午
21 夏至
28 十九

9 September

一	二	三	四	五	六	日
		1 廿五	2 廿六	3 廿七	4 廿八	5 廿九
6 三十	7 白露	8 初二	9 初三	10 教师节	11 初五	12 初六
13 初七	14 初八	15 初九	16 初十	17 十一	18 十二	19 十三
20 十四	21 中秋节	22 十六	23 秋分	24 十八	25 十九	26 二十
27 廿一	28 廿二	29 廿三	30 廿四			

10 October

一
4 廿八
11 初六
18 十三
25 二十

January

一	二	三	四	五	六	日
				1 元旦	2 十九	3 二十
4 廿一	5 小寒	6 廿三	7 廿四	8 廿五	9 廿六	10 廿七
11 廿八	12 廿九	13 腊月	14 初二	15 初三	16 初四	17 初五
18 初六	19 初七	20 大寒	21 初九	22 初十	23 十一	24 十二
25 十三	26 十四	27 十五	28 十六	29 十七	30 十八	31 十九

风从海上来

Wind from the Sea

中国，历史悠久，物华天宝，光辉灿烂，自古便吸引着世界的目光。

中国的北部与西部地区，或崇山环绕，或黄沙漫漫，在交通不发达的古代成为阻断往来的天堑。而富庶的东部地区，有天然良港，有每年如期而至的信风，更重要的是有许多可供歇脚补给的岛屿，于是最早的对外交流通道便在海上诞生。这条通道由于以大宗丝绸的出口为主，被称为“海上丝绸之路”。

广州是海上丝绸之路的东方发祥地，黄埔古港是最早的始发港。自秦汉时期起，从这里扬帆的船只远航至东南亚、南亚甚至更远的非洲地区，不仅传递财富，也传递文化，点亮了中外交流的天空。

China, esteemed as the country in the center of heaven, owning a large sum of precious resources, is splendid and brilliant with a long history. Since the ancient time, China has attracted the worldwide attention.

Surrounded by precipitous mountains and stretching deserts, the northern and western areas of China were natural barriers that prevented communication in the ancient time when transportation was poorly developed. While in the rich eastern areas, the natural good harbours, the regular annual monsoons, and more importantly, a considerable number of islands that could be used for provision and rest, gave birth to the earliest maritime transporting route. Mainly used for exporting silk, a bulk stock exchange, this route was known as the “Maritime Silk Route”.

The cradle of the Maritime Silk Route in the East was Canton, and the Whampoa port was the earliest starting point. Since the Qin and Han dynasties, the ships which departed from here and reached as far as South-eastern Asia, Southern

January / 2021.1

1

星期五

Friday

庚子鼠年 戊子月

十一月十八

元旦

New Year's Day

Asia and even the African, not only exchanged fortunes and cultures but also shed light on the communication between China and the outside world.

清咸丰广彩人物花蝶纹花盆托。

Flowerpot tray with patterns of figures, flowers and butterflies, reign of Emperor of Xianfeng, Qing dynasty.

January / 2021.1

2

星期六

Saturday

庚子鼠年 戊子月

十一月十九

January / 2021.1

3

星期日

Sunday

庚子鼠年 戊子月

十一月二十

清代象牙镂雕花瓣形人物花篮提梁带盖果盒。此提盒通体精镂细刻，刀工纯熟精炼，线条柔和明快，是清代广式象牙工艺的精品。汉晋以后的古籍中有许多关于广东产象的记载，而岭南的象牙制作甚至可以追溯到新石器时代晚期。

Ivory flower-shaped covered box with pierced patterns of willows, pavilions, figures and flowers, Qing dynasty. The delicately carved box, with its proficient and refined carving technique, and bright and soft lines, is a masterpiece of the Cantonese ivory crafts in the Qing dynasty. There are many records about the production of carved ivory in Guangdong in the ancient books written after the Han and Jin dynasties. And the ivory objects made in Lingnan region can even be traced back to the late Neolithic period.

岭南断交通

The Barricade of the Transport in Lingnan Region

在以农为本的中国古代社会，岭南地区一直都是个性鲜明的存在。一方面，作为中央王朝的治下之地，它一直积极吸纳以儒家为代表的中原文化，使得岭南人有着深厚的乡土观念；另一方面，岭南人又具有农耕民族少见的外向开拓和冒险精神。而岭南呈现出二元文化属性的原因，则在于其向内陆封闭、向海开阔的独特地理格局。

岭南通常指五岭以南的地区。绵延在岭南北部的五岭，是长江和珠江两大流域的分水岭，也是华中和华南气候的分界线。这组山脉的存在，阻隔了岭南地区与中原的交通及经济文化联系——尽管自秦代起陆续开辟了 8 条翻越南岭的大通道，以及 200 多条小驿道，使中原的文化技术得以渗入，但岭南群山仍旧使岭南地区保留着富有特色的本土文化。

In the Chinese ancient society which was based on agriculture, the Lingnan region had always been an existence with distinguishing features. On the one hand, under the reign of the central imperial court, it assimilated the culture from the Central Plains represented by Confucianism, and thus had a deep spiritual connection with the motherland; on the other hand, the Lingnan residents also had the adventurous spirit for exploring, a spirit rarely found among the agricultural minorities. What accounts for Lingnan's plural cultural characteristics is the unique geography which is barred from the interior but open toward the ocean.

Lingnan usually refers to the region on the south of the five ridges, which stretch along the northern part of Lingnan, forming the watershed between the Yangtse River and the Pearl River, as well as the demarcation between Central China and Northern China. Lingnan still maintained its own local cultural characters, because of the ridges

January / 2021.1

4

星期一

Monday

庚子鼠年 戊子月

十一月廿一

that hindered the transport, as well as economic and cultural connection between Lingnan region and the Central Plains–though since the Qin dynasty, 8 passages crossing Lingnan region and more than 200 small post roads had been built, allowing the cultures and techniques from the Central Plains to permeate into the area.

约绘于 1770 年的《广州鸟瞰图》，呈现了古广州城倚江而建的城市格局，越秀山城垣、观音阁、镇海楼都清晰可见。

Canton seen from a bird's-eye view, painted around 1770. It depicts the city layout of Canton which is constructed near the river; the city wall on Yuexiu Mountain, Pavilion of Bodhisattva and Zhenhai Pagoda can be seen in the painting.

小寒

Slight Cold

January / 2021.1

5

星期二

Tuesday

庚子鼠年 己丑月

十一月廿二

八门拓海路

Eight Estuaries

岭南地区的东部和南部濒临广阔的海洋，珠江尾闾的辫状水系不仅在广东中南部堆积出约5.6万平方公里的三角洲，还在东部沿海地区形成了8个入海口。它们有着成为良港的自然条件，仿佛陆地通向海洋的门户，因此人们以“门”字为它们命名——虎门、蕉门、洪奇门（沥）、横门、磨刀门、鸡啼门、虎跳门和崖门。

早在四五千年前，岭南先民就已经将目光从南海之滨投向更广阔的海域。他们驾着平底小舟，追逐着鱼群，并在南海乃至南太平洋沿岸及其岛屿形成了以陶器为纽带的贸易交换圈，为岭南地域文化添上了一笔浓厚的海外贸易色彩，也塑造了岭南人开放包容、敢为天下先的精神。这样的文化属性，也使后来“海上丝绸之路”的生发和十三行的兴起变得顺理成章。

The eastern and southern parts of Lingnan region face toward the vast ocean. The braided drainage in the downstream of the Pearl River not only heaped up a delta of 56,000 square kilometers in the central and southern part of Guangdong, but also formed 8 estuaries on the eastern coastal area. With the natural conditions capable of developing good harbours, the estuaries seemed to be the entrance leading to the ocean, and hence were deemed as the “gates” (in Pinyin “Men”), they are called Hu Men (Bocca Tigris), Jiao Men, Hongqi Men, Heng Men, Modao Men, Jiti Men, Hutiao Men and Ya Men.

Earlier four or five thousands years ago, the ancestors of Lingnan already shifted their attention from South China Sea to the wilder ocean. Steering the small boats to fish, they developed trading hubs that mainly exchanged porcelain among South China Sea, the coasts of the southern Pacific Ocean and other islands. Thus, ☞

January / 2021.1

6

星期三

Wednesday

庚子鼠年 己丑月

十一月廿三

a characteristic of foreign trade was deeply added into the local Lingnan culture, shaping the Lingnan people's dispositions of open-mindedness, tolerance, daring and adventurous spirits. With such a cultural characteristic, the "Maritime Silk Route" would come into being and the Thirteen Hongs in the later period would thrive in due course.

《远眺虎门港》。作为岭南东部沿海地区的8个入海口之一，虎门是得天独厚的天然良港。（图片采自《历史绘画》，香港艺术博物馆藏品）

The view of Bocca Tigris. As one of the eight estuaries on the eastern coastal area in Lingnan region, Bocca Tigris is a port that occupies a natural vantage point. (The picture is taken from *The Historical Pictures*, Collection of the Hong Kong Museum of Art.)

January / 2021.1

7

星期四

Thursday

庚子鼠年 己丑月

十一月廿四

造化钟广州

Canton's Fortune Endowed by Nature

今日的广州远离海洋，怎么会成为享誉世界两千多年的贸易大港呢？这，要归功于自然。在自然之手的造化下，广州一直在变，又始终未变。广州地处珠江口北部，六千年前这里还是一片汪洋大海，白云、越秀等山脉仅是海中的点点浮丘。伴随着西江、北江和东江等多条河流的共同堆积作用，到了两千多年前，这里出现了一片将白云山和越秀山连在一起的陆地，陆地上还有一座名叫南武城的小小城池——这就是广州早期的模样。

当时越秀山以南依然是大片海水，与珠江连成一片。每到涨潮的时候，浪花直拍越秀山脚。今天广州城内地势低平的越秀、荔湾、海珠、天河等区域，大部分都还沉在海底。随着时间的推移，它们才自水中慢慢显露出来，勾勒出广州的轮廓。

Today's Canton is far away from the ocean, then why it had become a major trading port renowned in the world for more than two thousand year? This owed to its natural condition. Canton has been changing but still remains the same with the endowment of nature. Canton is located on the north of the estuary of the Pearl River, which was inundated by the ocean six thousand years ago, when Baiyun and Yuexiu ridges were no more than small islands above the sea. Two thousand years ago, due to the common accumulation of the West River, North River and East River, a land that connected Baiyun Mountain and Yuexiu Mountain appeared. A small city called Nanwu City (Southern Martial City) also came into being–such was the condition of Canton in the early period.

At that time, the part on the south of Yuexiu Mountain was still swamped by the ocean and converged with the Pearl River. With the tide rising, the waves patted

January / 2021.1

8

星期五

Friday

庚子鼠年 己丑月

十一月廿五

the foot of Yuexiu Mountain. Some districts in today's Canton, such as Yuexiu, Liwan, Haizhu and Tianhe on the lower terrain, were then mostly situated under water. As time passed by, they gradually appeared above the water and formed the contour of Canton.

January / 2021.1

9

星期六

Saturday

庚子鼠年 己丑月

十一月廿六

清光绪广彩开光花鸟人物纹双象耳洗口瓶。

Vase with handles and patterns of figures, flowers and birds, reign of Emperor of Guangxu, Qing dynasty.

清光绪广彩描金花鸟人物纹带盖执壶。

Kettle with gold-painted patterns of figures, flowers and birds, reign of Emperor of Guangxu, Qing dynasty.

January / 2021.1

10

星期日

Sunday

庚子鼠年 己丑月

十一月廿七

海贸扬声名

The Renowned Maritime Trade

史学家普遍认为广州城古称番禺城，这一称谓源自《山海经》中记载的“贲隅”。根据明代黄佐所著《广东通志》记载，公元前862年，楚国已在广州这块风水宝地建城，名为楚庭郢。400年后，被越灭国的吴人迁入岭南，在楚庭的基础上建造了吴南武城，大约一百多年后，又被越国占领，史称越南武城。

尽管目前考古界还没有发现能证明南武城确实存在的有力物证，但在先秦时期，对于内陆各诸侯国而言，南武城之名可谓如雷贯耳。从岭外迁移而至的早期“外来户”，把岭南先民留下来的海上陶器贸易交换圈全盘继承，并进一步发扬光大，增加了很多贸易交换品类，其中不乏对中原地区而言相当珍贵稀有的宝贝。

It is universally acknowledged by the historians that, in the ancient time, Canton was called as Panyu, a name originated in *The Classic of Mountains and Seas*. According to *The General History of Guangdong*, written by Huang Zuo, in 862 B.C., the Chu State founded a city in Canton, a place with abundant natural resources, and the city was called Chutingying (the City of the Chu court). Four hundred years later, the people from the Wu State, which was destroyed by the Yue State, migrated to Lingnan region and founded Nanwu City (Wu's Southern Martial City) on the foundation of Chutingying. About one hundred year's later, the city was again occupied by the Yue State and named by the historians as Yuenanwu City (Yue's Southern Martial City).

The archaeologists have not found any convincing evidence that proves the existence of Nanwu City, but in the pre-Qin dynasty, for the dukes and princes of the feudal states, the name of Nanwu City was as sonorous as a thunder. The early ☞

January / 2021.1

11

星期一

Monday

庚子鼠年 己丑月

十一月廿八

immigrants that came to Lingnan region inherited all the maritime trading hubs of pottery wares from the ancestors of Lingnan, and developed them by adding more trading commodities, among which there were precious objects rarely seen in the area of the Central Plains.

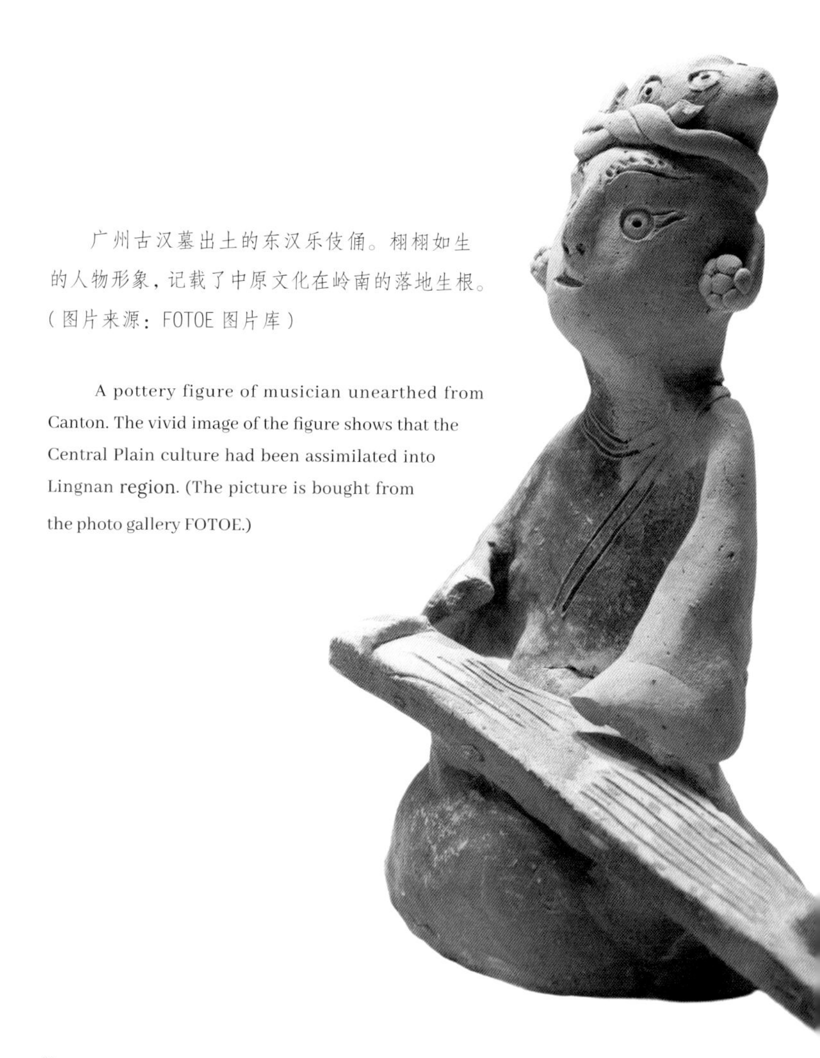

广州古汉墓出土的东汉乐伎俑。栩栩如生的人物形象，记载了中原文化在岭南的落地生根。（图片来源：FOTOE 图片库）

A pottery figure of musician unearthed from Canton. The vivid image of the figure shows that the Central Plain culture had been assimilated into Lingnan region. (The picture is bought from the photo gallery FOTOE.)

January / 2021.1

12

星期二

Tuesday

庚子鼠年 己丑月

十一月廿九

豪杰铸伟城

A City Built by the Outstanding Generals

南武城逐渐发展为繁荣的对外贸易商港，但与之形成鲜明对比的是势力日渐衰落的越国。为了得到庇护，越国经常向魏国进献大量礼物，如犀角、象牙、翡翠、珠玑等，这些珍宝皆来自于南武城的对外贸易。南武城令人垂涎的财富引起了秦始皇的觊觎。《竹书纪年》里提到，控制对外贸易权，获取源源不断自海外而来的珍稀货物，正是秦始皇经略岭南的目的之一。

公元前 214 年，秦始皇命任嚣、赵佗率领秦军完成平定岭南大业。整个岭南由此被划入秦朝的版图，设南海郡、桂林郡、象郡三郡，任嚣被任命为南海郡尉。南海郡相当于今广东省的大部分地区。鉴于番禺北有山岭为屏、南与江海相通的优越地理位置，此地被定为南海郡的郡治。据史料记载，任嚣扩建了南武城，更名为任嚣城，广州的建城史由此开始。

As Nanwu City gradually developed to be a prosperous trading port with the outside world, there came with a sharp contrast that the power of the Yue State was declining. In order to get protection, the Yue State often paid a large amount of tribute to the Wei State, such as rhino horns, ivory, jades and beads which were all precious objects from foreign trade in Nanwu City. Then Qinshihuang (the First Emperor of the Qin State) cast his covetous eyes on the alluring fortune of Nanwu City. *The Bamboo Book of Chronology* mentioned that Qinshihuang's aim was to control the foreign trade in order to acquire the inexhaustible rare commodities from abroad by including Lingnan region into his territory.

In 214 B.C., Qinshihuang entrusted two generals, Ren Xiao and Zhao Tuo, with the great cause of leading the Qin army to conquer Lingnan region. Therefore, the whole region of Lingnan was included into the territory of the Qin dynasty, and ☞

January / 2021.1

13

星期三

Wednesday

庚子鼠年 己丑月

十二月初一

was divided into Nanhai County, Guilin County and Xiangjun County. Ren Xiao was appointed as the captain of Nanhai County which occupied most of today's Guangdong Province. In view of the superior geographical position of Panyu with mountains on the north and rivers connecting to the sea on the south, Panyu was designated as the capital of Nanhai County. According to historical records, Ren Xiao expanded Nanwu City and renamed it as City of Ren Xiao. Thus the history of Canton began.

在广州惠福东路出土的南越国水闸遗址。水闸位于当时的珠江北岸，南北向，闸口宽5米，南北长35米，闸口的南北均呈“八”字形敞开。（图片来源：FOTOE 图片库）

The site of the water gate of the Nanyue Kingdom. The water gate was located on the northern shore of Pearl River, running from the south to the north, with a width of 5 meters and length of 35 meters; both the southern and northern part of the water gate were flared, resembling the shape of the Chinese character "eight" . (The picture is bought from the photo gallery FOTOE.)

January / 2021.1

14

星期四

Thursday

庚子鼠年 己丑月

十二月初二

南越盛风华

The Nanyue Kingdom

对于秦始皇的意图，任嚣和赵佗自然心领神会，尤其当他们在岭南地区深耕数年之后，对这里的区位优势认识得更加深刻。秦末农民起义，天下大乱，赵佗遵任嚣的遗令，封关、绝道，筑起三道防线，拥兵数十万开创南越国，自立为王。

由于北向的通道被赵佗主动断绝，中原物资无法流入，赵佗便一面开发岭南，一面积极开拓海外贸易通道，为“海上丝绸之路”的形成奠定了基础。当时南越国已经可以造出宽 8 米、长 30 米、载重五六十吨的木船。每年冬季，满载着丝织物、布匹、漆器、陶器和青铜器的商船乘着东北季风向南到印度洋，再顺着夏季的西南季风运回犀角、象牙、珍珠、玳瑁、玻璃和金银器等。

Ren Xiao and Zhao Tuo obviously understood Qinshihuang's intention. Especially after they had administrated Lingnan region for several years, they had a better understanding of the geographical advantages here. At the end of the Qin dynasty, when the peasant revolt caused chaos all over the country, Zhao Tuo followed Ren Xiao's will, closed the passes and blocked down the routes, built three defensive lines, established the Nanyue Kingdom with hundreds of thousands of soldiers and crowned himself as the king.

Because the northern channel was cut off by Zhao Tuo, materials could not be imported from the Central Plains, Zhao Tuo continued to develop Lingnan region while explored new maritime trading routes, which laid the foundation for the formation of the "Maritime Silk Route". At that time, the Nanyue Kingdom had been able to build wooden ships with a width of 8 meters, a length of 30 meters and a load of 50 to 60 tons. Every winter, the merchant ships full of silk fabrics, cloth, lacquer

January / 2021.1

15

星期五

Friday

庚子鼠年 己丑月

十二月初三

wares, pottery and bronze sailed southward to the Indian Ocean with the northeast monsoon, and went back with rhinoceros horns, ivory, pearls, tortoiseshells, glass, gold and silver wares with the southwest monsoon in summer.

January / 2021.1

16

星期六

Saturday

庚子鼠年 己丑月

十二月初四

19 世纪玻璃画仕女图。玻璃画是外销画的重要画种之一，制作者采用反笔彩绘技法，在玻璃背面施以油彩绘制。这幅画色彩亮丽、笔触细腻。画中的仕女手捧茶杯，充满浓厚的东方情调。

A maiden, reverse glass painting, 19th century. Reverse glass paintings are one of the most important export paintings. The artisans applied the reverse painting technique to paint on the back of the glass. This painting is bright in color and delicate in strokes. In the painting, the lady holding the tea cup is characterized by a strong oriental sentiment.

January / 2021.1

17

星期日

Sunday

庚子鼠年 己丑月

十二月初五

晚清时期的菩提叶水彩人物画。

A figure on bodhi leaf, watercolour, late Qing dynasty.

丝绸通海陆

The Silk Route Connecting the Land and the Sea

如果说汉之前的海上贸易只是区域范围的“小动作”，那么在公元前 111 年汉武帝灭南越，重新将岭南地区置于中央王朝统治之下后，以广州为中心的海上贸易首次将遥远的印度洋与广阔的中国内陆连接起来，一条突破区域局限、影响全国的陆海贸易大通道——“海上丝绸之路”出现了。

中原的各种货物和特产畅通无阻地自水路、陆路和海路进入岭南，再从广州、徐闻、合浦等港口转运到波斯、斯里兰卡，乃至古罗马帝国。广州亦从岭南中心城市一跃成为全国九大都会之一。司马迁在《史记·货殖列传》中写道：“番禺亦一都会也，珠玑、犀、玳瑁、果、布之凑。”班固在《汉书》中则有更详细的描述：“处近海，多犀、象、毒冒、珠玑、银、铜、果、布之凑，中国往商贾者，多取富焉。番禺，其一都会也。”两千多年前，广州便是缔造传奇与财富的地方。

If the maritime trade before the Han dynasty promoted the interaction limited in the regional scope, then after Hanwudi (the Martial Emperor of the Han dynasty) destroyed the Nanyue Kingdom in 111 B.C. and put Lingnan region under the central court's reign again, the maritime trade, with Canton as its center, connected the remote Indian Ocean with the vast mainland of China for the first time. Hence a huge maritime route——the “Maritime Silk Route” which broke through the regional limitation and affected the whole country's trade both on the land and ocean appeared.

All kinds of goods and specialties in the Central Plains were transported into Lingnan region by waterways, overland routes and sea routes, and then shipped to Persia, Sri Lanka and even the ancient Roman Empire through the ports such as Canton, Xuwen, Hepu and so on. Canton had also become one of the nine largest ☞

January / 2021.1

18

星期一

Monday

庚子鼠年 己丑月

十二月初六

cities in China. Sima Qian (a famous Chinese historian) wrote in *The Historical Records·Inventory of Foreign Goods*: "Panyu is also the capital. The beads, rhino horns, tortoiseshells, fruits and cloth are all gathered here." Ban Gu (a historian in the Eastern Han dynasty) gave a more detailed description in *The Book of Han Dynasty*: "in the coastal areas, there are many rhino horns, ivory, tortoise shells, beads, silver, copper, fruits and cloth. The Chinese merchants who go there to trade often become rich. Panyu is such a metropolis." More than 2000 years ago, Canton was already the place to create legends and wealth.

在广州番禺东汉古墓群出土的彩绘骑马俑。（图片来源：FOTOE 图片库）

Colourful pottery figure of a cavalryman, unearthed in the ancient tomb of the Eastern Han dynasty in Panyu, Canton. (The picture is bought from the photo gallery FOTOE.)

January / 2021.1

19

星期二

Tuesday

庚子鼠年 己丑月

十二月初七

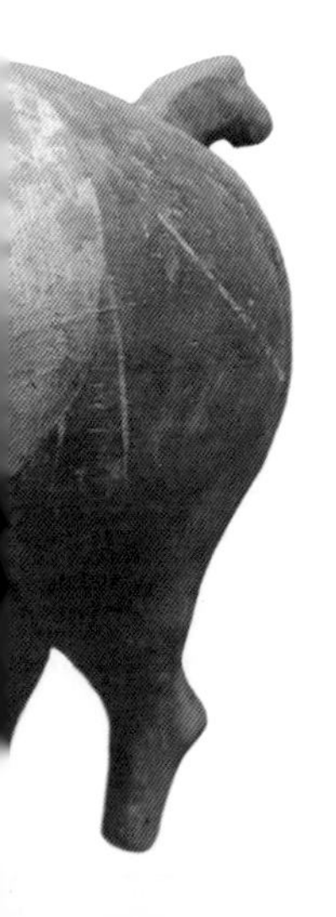

海客乘风至

Foreign Guests Arrived with the Monsoon

商品贸易其实是另一种形式的文化交流，特别是对外贸易。那些精心挑选的货品，不经意间展现着一个国家或地区的技术和美学高度；而跟随商船与货物而来的人，也同样通过自己的言行举止，有意无意地传播着本国的文化。这些跟随西南季风而来的人，多数是借着使节之名为利奔波的商人，还有一部分是商人的随行仆从、慕名前来学习中华文化的人，以及肩负着传播宗教福音使命的僧侣。在以千年计算的岁月里，他们仿佛百流归海，不断地丰富着岭南文化的内涵，塑造着广州这座城市的人文气质。

秦汉时的番禺已出现外国人的身影，广州汉墓出土的20余件胡人俑座灯十分耐人寻味。这些本土制造的座灯上有着栩栩如生的胡人形象，表明汉代的番禺已活跃着不少来自东南亚、南亚、非洲及西域各国的胡人，从事奴仆或其他工作。

The commercial trade, especially the foreign trade, is in fact another form of cultural exchange. Those carefully selected goods inadvertently showed the degree of technology and aesthetics of a certain country or region, while those who came with the merchant ships and goods also unconsciously spread their own culture by their own words and deeds. Most of the people were merchants who sailed with the southwest monsoon to arrive here to seek for fortune and fame while disguised themselves as envoys. Some of them were the merchants' servants, and some of them were those who were attracted by Chinese culture and came to learn, and some were the monks who had the mission of spreading religious gospel. Throughout thousands of years, they were like hundreds of flows that assembled into the ocean, constantly enriching the connotation of Lingnan culture and shaping the city's cultural characters.

☞

大寒

Great Cold

January / 2021.1

20

星期三

Wednesday

庚子鼠年 己丑月

十二月初八

腊八节

Laba Festival

Foreigners appeared in Panyu as early as in the Qin and Han dynasties. It is very interesting that more than 20 terracotta lamps made in the shapes of foreign people have been unearthed in the Tombs of the Han dynasty in Canton. The lamps, on which there are vivid figures of the foreigners, were made in the local area, indicating that many foreigners from Southeast Asia, South Asia, Africa and Western regions had been arrived in Panyu in the Han dynasty, and they were engaged actively in servitude or other work.

广州汉墓出土的胡人俑陶座灯。（图片来源：FOTOE 图片库）

The pottery figure of a foreigner unearthed in the Han Tomb in Canton. (The picture is bought from the photo gallery FOTOE.)

January / 2021.1

21

星期四

Thursday

庚子鼠年　己丑月

十二月初九

大港数羊城

Canton, the Grand Port

隋唐时期，由于西域连年战争，“陆上丝绸之路”被阻断，“海上丝绸之路”再次迎来辉煌，广州成为四大港口（广州、泉州、明州、扬州）之一，并与扬州、汴梁并称全国三大商业城市，是世界著名的东方港市。每年从海外来此做生意的商船多达4000余艘。一到舶期，每日可征得高达15万两白银的税收（以清末白银计）。由于往来朝贡经商的商船实在太多，唐高宗显庆六年（公元661年），广州首设市舶使，由宦官担任，总管海路邦交外贸，是中国最早的对外贸易管理职位，也是宋、元、明三朝一脉相沿的市舶司前身。当时市舶使的主要职责有四项：一是向前来贸易的船舶征收关税；二是代表宫廷采购一定数量的舶来品；三是管理商人向皇帝进贡的物品；四是对市舶贸易进行监督和管理。

During the Sui and Tang dynasties, due to the successive years of wars in the western regions, the “Overland Silk Road” was blocked, and the “Maritime Silk Route” once again prospered. Canton became one of the four major ports (Canton, Quanzhou, Mingzhou, Yangzhou), and was also known as the three major commercial cities of China together with Yangzhou and Bianliang. It was a famous oriental port in the world. Every year, more than 4,000 merchant ships came here. Once they arrived at the port, they were collected to a daily tax of up to 150,000 tael of silver (counted as silver at the end of the Qing dynasty). Because there were so many merchant ships coming and going to pay tribute and trade, in the sixth year of Tanggaozong’s reign in the Tang dynasty (661 A.D.), Canton was the first one to set up the position of Municipal Shipping Administrator, taken by an eunuch to be in charge of maritime routes, foreign affairs and trade. It was the earliest foreign

January / 2021.1

22

星期五

Friday

庚子鼠年 己丑月

十二月初十

trade administrating position in China, and also the predecessor of the Municipal Ship Department in the Song, Yuan and Ming dynasties. At that time, the main responsibilities of the Municipal Shipping Administrator were as follows: first, collecting customs duties on the ships coming forward to trade; second, purchasing a certain number of imported goods on behalf of the court; third, managing the goods that the merchants paid tribute to the emperor; fourth, supervising and managing the city's shipping and trade.

清道光广彩花蝶人物纹马克杯。

Mug with patterns of figures, flowers and birds， reign of Emperor of Daoguang, Qing dynasty.

January / 2021.1

23

星期六

Saturday

庚子鼠年 己丑月

十二月十一

清光绪广彩花鸟人物纹小花盆。

Small flowerpot with patterns of figures, flowers and birds, reign of Emperor of Guangxu, Qing dynasty.

January / 2021.1

24

星期日

Sunday

庚子鼠年 己丑月

十二月十二

风华耀海丝

The Marvelous Maritime Silk Route

有别于明清时期的“闭关锁国”，唐朝很欢迎外国人的到来。据《新唐书·地理志》记载，当时进入大唐疆域的通道一共有七条：“一曰营州入安东道，二曰登州海行入高丽渤海道，三曰夏州塞外通大同云中道，四曰中受降城入回鹘道，五曰安西入西域道，六曰安南通天竺道，七曰广州通海夷道。”

“广州通海夷道”就是唐宋时期的“海上丝绸之路”，也是当时世界上最长的航线。由于“陆上丝绸之路”被战争阻断，加上这时的造船术和航海术有了长足发展，远洋航线不断被拓展与延伸。它以广州为起点，长达 1.4 万公里，从今香港屯门港驶向外海，经东南亚、印度洋北部诸国、红海沿岸、东北非和波斯湾诸国等 30 余国，抵达奥巴拉港和巴士拉港，再换乘小船沿幼发拉底河辗转至阿拉伯帝国首都（今巴格达）。

Different from the "closed-door policy" in the Ming and Qing dynasties, the Tang dynasty welcomed the arrival of foreigners. According to *The New Book of Tang Dynasty·Geography Records*, at that time, there were seven channels to enter the territory of China in the Tang dynasty: “the first one is from Yingzhou to Andong, the second is from Dengzhou to Gaoli through a sea route, the third is from the outer territory Xiazhou to Datong, the fourth is from the central Shouxiang city to Huihu, the fifth is from Anxi to the West, the sixth is from Annan to India, the seventh is the maritime route leading to Canton.”

“The maritime route to Canton” was the "Maritime Silk Route" in the Tang and Song dynasties, and it was also the longest route in the world at that time. Because the "Overland Silk Road" was blocked by war, along with the rapid development of ship-building technology and navigation at that time, the ocean route was ☞

January / 2021.1

25

星期一

Monday

庚子鼠年 己丑月

十二月十三

continuously expanded and extended. It started from Canton and stretched 14,000 kilometers. Departing from today's Tuen Mun port in Hong Kong to the open seas, it passed through more than 30 countries, including Southeast Asia, several countries on the northern Indian Ocean, Red Sea coast, northeastern part of Africa and the Persian Gulf, then arrived at the Obala and Basra ports, and finally turned to Baghdad, the capital of the Arabian Empire, by using small boats along the Euphrates River.

17 世纪荷兰画家的画作，内容为在珠江眺望广州城。早在唐宋时期，以广州为起点的“海上丝绸之路”，已长达 1.4 万千米。

The Landscape of Canton by the Side of Pearl River was painted by a Dutch painter in the 17th century. As early as the Tang and Song dynasties, the "Maritime Silk Route" starting from Canton, had already been 14,000 kilometers long.

January / 2021.1

26

星期二

Tuesday

庚子鼠年 己丑月

十二月十四

番坊聚五方

The Foreigners' Quarter

唐代鉴真和尚曾描绘他在广州亲见的景象："江中有婆罗门、波斯、昆仑等大舶不知其数，并载香料、珍宝积载如山。舶六七丈。狮子国、大食国、骨唐国、白蛮、赤蛮等往来居住，种类极多。"

在众多前来贸易的外国商人中，不乏在广州落地生根的。早在南北朝时期，就有阿拉伯商人居留广州。据史料记载，唐代广州番人约达 12 万。唐政府对外国人管制颇为严厉，不许他们随意在城内活动，更担忧华夷杂处，容易勾结作乱，因此在广州城以西临近珠江之处辟出番坊供番商居住，范围大抵为今北到中山路，南达惠福路和大德路，西抵人民路，东达解放路。番坊中居住的，大部分是信仰伊斯兰教的阿拉伯人。他们在番坊建造了清真寺——怀圣寺，寺中建有光塔，光塔通体光滑洁白，久之被当作引航灯塔。

Jian Zhen, a famous monk in the Tang dynasty, once described the scene of Canton in his eyes: "on the river there are countless big ships from India, Persia, Southeast Asia and other places; the spices, and treasures on the ships are piled up like mountains. The ships are as long as six to seven feet. There are many people from various countries such as Shizi state (ancient Sri Lanka), Dashi state (the ancient Arabian Empire), Gutang state (an ancient country presently unknown to us), Baiman (which probably referred to the West) and Chiman (which probably referred to Africa and Arab.)"

Many foreign traders who came here had taken root in Canton. As early as the Northern and Southern dynasties, there were Arabian traders who lived in Canton. According to some historical records, around 120,000 foreigners lived in Canton in the Tang dynasty. The government of the Tang dynasty was afraid that if the local ☞

January / 2021.1

27

星期三

Wednesday

庚子鼠年 己丑月

十二月十五

people and foreigners got together, they might collude with each other and result in revolt, so rigid control over foreigners was implemented. Foreigners were not allowed to wander inside the city freely and thus, Fanfang (the Foreigners' Quarter) was built near the Pearl River on the west of Canton for foreign merchants to live in. The range was from Zhongshan Road in the north, HuiFu Road and Dade Road in the south, Renmin Road in the west and Jiefang Road in the east. Most of the people living in Fanfang were Arabians who believed in Islamism. They built a mosque in Fanfang——Huaisheng Temple, in which there was a minaret, which was smooth and white, and used as a beacon for a long time.

清代时期，在广州街市上采办商品的瑞典商人。（图片采自《昔日乡情》）

The Swedish merchant in a Cantonese market in the Qing dynasty. (The picture is taken from *The Nostalgia for the Hometown*.)

January / 2021.1

28

星期四

Thursday

庚子鼠年 己丑月

十二月十六

小海抱坡山

The Small Sea Embracing the Po Hill

作为当时世界上最著名的港口之一，唐代广州的港口建设和规划也值得一提。当时广州城已经发展出内港码头和外港码头进行分流。内港有坡山码头和兰湖码头（今流花湖一带）。坡山码头即今惠福西路五仙观所在地。汉唐之际，坡山是一座海拔约 20 米的丘陵，其西为珠江，由于水面太过浩荡，当时被称为“珠海”，也叫“小海”，与“大海”狮子洋相对。坡山码头由怀圣寺光塔引航，主要用来停泊外国商船。

As one of the most famous ports in the world back then, the construction and planning of the ports in Canton in the Tang dynasty is also worth to introduce. At that time, Canton had developed internal and external ports for diversion. There were Poshan Wharf and Lanhu Wharf in the inner port (now near Liuhua Lake). Poshan wharf is now the location of Wuxianguan, which is on HuiFu West Road. In the Han and Tang dynasties, Poshan Hill was a hill with an altitude of about 20 meters, to the west of which, was the Pearl River. Because the water surface was vast and wavy, it was called "Zhuhai" (the Sea of Pearl) or "small sea" at that time, forming a contrast to the “big sea” which referred to the shizi Ocean (the estuary of the Pearl River). Poshan wharf, mainly used as the berth of foreign merchant ships, was piloted by the beacon, the minaret in Huaisheng Temple.

January / 2021.1

29

星期五

Friday

庚子鼠年 己丑月

十二月十七

清嘉庆广彩锦地山水仕女碟盘。

Plate with patterns of maidens and landscape on brocade background, reign of Emperor of Jiaqing, Qing dynasty.

January / 2021.1

30

星期六

Saturday

庚子鼠年 己丑月

十二月十八

January / 2021.1

31

星期日

Sunday

庚子鼠年 己丑月

十二月十九

清乾隆广彩花卉徽章纹盖瓶。

Vase with lid with patterns of flowers and armorial, reign of Emperor of Qianlong, Qing dynasty.

February

一	二	三	四	五	六	日
1 二十	2 廿一	3 立春	4 小年	5 廿四	6 廿五	7 廿六
8 廿七	9 廿八	10 廿九	11 除夕	12 春节	13 初二	14 初三
15 初四	16 初五	17 初六	18 雨水	19 初八	20 初九	21 初十
22 十一	23 十二	24 十三	25 十四	26 元宵节	27 十六	28 十七

波罗震南海

The God of South China Sea

除了内港码头，当时外港码头也有两个，分别是今香港新界的屯门码头和黄埔庙头村的扶胥码头。从广州驶离的商船，在两个外港码头做最后的停留补充之后，便要驶向茫茫大海。

扶胥码头位于东江与珠江汇流入海处，是古代商船进出广州的必经之处。公元 594 年，隋文帝下诏在扶胥码头附近修建南海神庙，这意味着广州作为第一对外贸易大港的地位得到了官方的认可。自唐以来，各朝政府均设有专人管理庙事，历代皇帝都派官员到南海神庙举行祭典，留下了不少珍贵碑刻，故有“南方碑林”之称。南海神庙至今尚存，已有 1400 余年历史，每年农历二月广州人都会在此举行庆贺南海海神生日的庙会，称为“南海神诞”或“波罗诞”，是古代广州发达的海上贸易交通的一段生动注脚。

Besides the inner wharf, there were two outer wharves at that time, namely, Tuen Mun Wharf in the New Territories of Hong Kong and Fuxu Wharf in Miaotou village of Whampoa. The merchant ships departed from Canton, after making the final stopover for provision at these two outer ports, and then sailed to the boundless ocean.

Fuxu wharf, located at the confluence of the East River and the Pearl River, was the only pass for ancient commercial ships to enter and leave Canton. In 594 A.D., Suiwendi (Emperor of the Sui dynasty) ordered to build the Nanhai Temple near Fuxu wharf, indicating that Canton's status as the first foreign trade port had been officially recognized. Since the Tang dynasty, the government of each dynasty had set up special personnel to manage the temple affairs. The emperors of all dynasties sent officials to the Nanhai Temple to hold ceremonies, leaving many precious ☞

February / 2021.2

1

星期一

Monday

庚子鼠年 己丑月

十二月二十

inscriptions, so it is known as the "Southern Forest of Stone Tablets". The Nanhai Temple still exists today, with a history of more than 1400 years. The temple fair celebrating the birth of God of Nanhai is held in February of every lunar calendar. It is called the "Birthday of God of Nanhai" or "Boluo's Birthday", a vivid footnote of the well developed maritime trade and transportation in ancient Canton.

创建于隋开皇十四年（594 年）的广州南海神庙，又称波罗庙，位于广州市黄埔区南岗镇庙头村。（图片来源：FOTOE 图片库）

Built in 594A.D., the fourteenth year of the reign of Suikaihuang, the first emperor of the Sui dynasty, the Nanhai Temple in Canton is also called "Boluo Temple", located in Miaotou village, Nangang town, Whampoa. (The picture is bought from the photo gallery FOTOE.)

February / 2021.2

2

星期二

Tuesday

庚子鼠年 己丑月

十二月廿一

三城开华府

Canton, an Enlarged City

唐之前，广州城是单城；到了唐代，随着经济的发展和人口结构的复杂化，需要各区域有更加清晰的功能划分，以便能够高效管理。因而，唐代将城南珠江新淤成的平原纳入，从南向北分为南城、子城和官城。南城濒江（岸线在今一德路、万福路一线），船只停靠于此，既是货物进出的码头，也是商业中心区；子城即步骘城，是主要的生活居住区域；官城为官衙机构所在地（今财政厅一带）。一条南北向的主干道从刺史署直通江边（即如今的北京路），江边有广阳馆，用来接待官员和使者；又有一条东西向的主干道与北京路相交，横穿全城（即如今的中山路）。当时，官员和使者自城南码头下船，沿着南北主干道穿过三重城，所见皆为繁华壮丽之区，尽显气象。

Before the Tang dynasty, Canton was a city without satellite towns; in the Tang dynasty, with the development of economy and the complexity of population structure, a clearer functional division of each region was needed to ensure the management efficiency. Therefore, in the Tang dynasty, the newly silted plain of the Pearl River on the south was included into the city, which was divided into the southern city, the sub-city and the official city from the south to the north part. The south part of the city, not only a wharf for cargoes to enter and leave the port, but also a commercial center, was adjacent to the river (the coastline on today's Yide Road and Wanfu Road), where ships berthed. The sub-city, namely, Buzhi city, was the main residential area. The so-called official city was the place in which the government offices were located (today's provincial finance department is also in this area). A main road, running from the north to the south, led directly to the riverside from the supervision office (it has now become Beijing road). On the ☞

立春

Beginning of Spring

February / 2021.2

3

星期三

Wednesday

庚子鼠年 庚寅月

十二月廿二

riverside, there was Guangyang Hall for receiving officials and envoys. Another main road, running from the east to the west, intersecting Beijing Road and crossing the whole city, is now called Zhongshan Road. At that time, officials and envoys disembarked from the wharf in the south part of the city and went through the city along the north-southern main road. All they saw were prosperous and grand areas with a magnificent atmosphere.

19世纪初期的黄埔港风光。（图片采自《东西汇流》，香港艺术馆藏品）

The Landscape of Whampoa port in the early 19th century. (The picture is taken from *East meets West*, Collection of Hong Kong Museum of Art)

February / 2021.2

4

星期四

Thursday

庚子鼠年 庚寅月

十二月廿三

小年

kitchen God's Day

八镇卫三城

Eight Satellite Towns

两宋时期城市的商业化程度很高，本就具有浓厚商业基因的广州在这个时代如鱼得水。尽管在南宋后期，广州第一大港的地位被泉州取代，但并不意味着广州的对外贸易就此衰落；相反，由于商贸的发达，广州城在五代南汉城的基础上进一步扩大了。这个时期珠江江面已显著缩小，东关和西关平原逐渐出露，因此公元 1068 年，广州先于子城东扩筑东城，5 年后又向西联通番坊形成西城，成三城并立之势，城市结构进一步复杂化。不仅如此，宋代工商业的发展、科学技术的进步和对珠三角的进一步开发，促使广州周边形成了 8 个卫星城镇，分别是大通镇（今花地大通滘口）、瑞石镇（今穗石）、平石镇（今谢村东南“胜石”坊附近）、猎德镇（今猎德村）、大水镇（今河村附近）、石门镇（今石门村）、白田镇（今西关丛桂里附近）、扶胥镇（今南海神庙庙头村）。

In the Song dynasty, the commercialization level of the cities was very high. Canton, with a strong commercial gene, was like a duck to water in this era. Although in the late Southern Song dynasty, the position of Canton as the largest port was replaced by Quanzhou, it did not mean that Canton's foreign trade declined; on the contrary, due to the development of commerce and trade, Canton expanded further on the basis of the Southern Han City in the Five Generations period, when the width of the Pearl River had been significantly reduced, and the plains of Dongguan (the Eastern Pass) and Xiguan (the Western Pass) gradually exposed. Therefore, in 1068 A.D., Canton expanded its eastern city on the eastern part of the sub-city, and five years later, it connected with Fanfang (the Foreigner's Quarter) to built the western city, and thus the urban structure was further complicated, forming a situation where three cities were independent on one another. Moreover, the

February / 2021.2

5

星期五

Friday

庚子鼠年 庚寅月

十二月廿四

development of industry and commerce, the progress of science and technology and the further development of the Pearl River Delta in the Song dynasty led to the formation of eight satellite towns around Canton, namely Datong town (now Datong Jiaokou in Huadi district), Ruishi town (now Suishi), Pingshi town (near “Shengshi” lane in the southeastern side of today's Xie village), Liede town (now Liede village), Dashui town (now near today's He village), Shimen town (now Shimen village), Baitian town (now near “Conggui” lane, Xiguan), Fuxu town (now Miaotou village, Temple of the God of Nanhai).

清乾隆广彩描金锦地开光花鸟人物纹菱形花插。

Flower receptacle with patterns of gardens and figures over a brocade ground, reign of Emperor of Qianlong, Qing dynasty.

February / 2021.2

6

星期六

Saturday

庚子鼠年 庚寅月

十二月廿五

February / 2021.2

7

星期日

Sunday

庚子鼠年 庚寅月

十二月廿六

清乾隆广彩青花人物纹双螭耳狮钮盖方口瓶。

Vase with square lid, lion-shaped knob, and blue and white figural patterns, reign of Emperor of Qianlong, Qing dynasty.

西村高仿瓷

Delicate Knockoff Porcelain in Western Village

宋代经广州销往海外的商品中，瓷器占据首要地位。由于海外需求量大，尽管广州本地没有烧制瓷器所需的瓷土、燃料，烧窑技术更无法与其他名窑相比，从事海上贸易的商人为了减少运输成本、获取更大利润，还是在广州设窑仿烧其他名窑的出品，这就是西村窑，位于今广州西村体育场内（增埗河东岸岗地上）。

西村窑的一大显著特点就是大量仿制当时的名窑产品，包括河北磁州窑、浙江越窑、陕西耀州窑、江西景德镇窑和吉州窑、福建建窑、湖南长沙窑等，几乎所有的名窑都可以在西村窑找到其仿品。1956 年于佛山发掘的一座龙窑遗址，出土了青釉、青白釉、酱釉、绿釉等各类釉色的瓷器，器类多达 40 余种，“高仿”数量巨大，足见其当时的生产规模。

Porcelain was the most important products among all the goods sold overseas through Canton in the Song dynasty. Due to a large amount of overseas demand, although there was no clay and fuel for firing porcelain in Canton, and the kiln technology was not as good as that of the other famous kilns, in order to reduce the transportation cost and obtain more profits, the merchants engaged in maritime trade still set up kilns in Canton to imitate the products from other famous kilns. Thus appeared the Western Village Kiln, which is located in today's Stadium of Xicun Village (now on the down land of the eastern bank of Zengbu river) in Canton.

A remarkable feature of the Xicun Village Kiln was to copy a large number of products from famous kilns at that time, such as Cizhou kiln in Hebei province, Yueyao kiln in Zhejiang province, Yaozhou Kiln in Shaanxi province, Jingdezhen kiln and Jizhou kiln in Jiangxi province, Jianyao kiln in Fujian province, Changsha kiln in ☞

February / 2021.2

8

星期一

Monday

庚子鼠年 庚寅月

十二月廿七

Hunan province, etc. Knockoff porcelain, which imitated nearly all the products of the famous kilns, could be found in the Xicun Village. In 1956, a site of dragon kiln (a certain kind of Chinese kilns) was excavated, which unearthed various kinds of porcelain, including celadon, green and white glaze, sauce glaze, green glaze and other types. There are more than 40 types of utensils, with a large number of "high quality knockoff" porcelain, showing the scale of production at that time.

广州西村窑出土的宋代青白釉凤头壶。（图片来源：FOTOE 图片库）

A green and white pot unearthed in Xicun Village in Canton in the Song dynasty. (The picture is bought from the photo gallery FOTOE.)

February / 2021.2

9

星期二

Tuesday

庚子鼠年 庚寅月

十二月廿八

水陆大通道

A Grand Passage Connecting the Land to the Ocean

为提高运输效率，元代重修了京杭大运河，打破隋唐运河东西走向格局，直接贯通中国南北方，并开辟海运航线，构建了一条水陆联动的物流大通道。

这条国际级贸易大通道的核心枢纽是元大都，向北通过“陆上丝绸之路”连接起中亚各国、印度、阿拉伯及欧洲西部；向东、向南通过漕运与海运将扬州、宁波、泉州、广州等港口城市连接起来，再通过“东海丝绸之路”连接高丽、日本和琉球群岛；通过“南海丝绸之路”连接东南亚诸国、印度半岛、阿拉伯半岛、非洲东海岸及马达加斯加。当这条通道被元朝彻底打通之后，处于大通道各节点的城市间物资流转变得更加快速和顺畅，带动了城市商业的发展。以广州为例，当时与之有贸易往来的国家和地区有 141 个，占了全国对外交往国家和地区总数的 64%。

In the Yuan dynasty, the Beijing-Hangzhou Grand Canal was built to improve the transporting efficiency, get rid of the structure of Sui and Tang canals running from the east to the west, directly connect the northern with southern parts of China, open up new shipping routes, and construct a large logistics channel connecting the land to the ocean.

The core hub of this international trade route was Yuandadu (an ancient metropolis in the Yuan dynasty), which, through the "Overland Silk Road", connected China northwards with central Asian countries, India, Arabia and Western Europe; Yangzhou, Ningbo, Quanzhou, Canton and other port cities in the East and South were connected by waterway channels and sea routes, and then linked to Koryo, Japan and Ryukyu Islands through the “East China Sea Silk Route”; while Silk Route of South China Sea linked the southeast Asian countries, Indian Peninsula, Arabian ☞

February / 2021.2

10

星期三

Wednesday

庚子鼠年 庚寅月

十二月廿九

Peninsula, the east coast of Africa and Madagascar. After the passage was completely opened by the Yuan dynasty, the logistics among various hubs became more efficient and smooth, which led to the commercial development in the cities. Take Canton for example, it had traded with 141 countries and regions at that time, accounting for 64% of the total number of the foreign countries and regions that traded with China.

明代画家仇英的《抗倭图卷》(节选)，生动呈现了倭寇对沿海居民的危害。

Fighting against the Japanese pirates(excerpt), by Chou Ying (a painter in the Ming dynasty), vividly shows that the Japanese pirates had done harm to the coastal residents.

February / 2021.2

11

星期四

Thursday

庚子鼠年 庚寅月

十二月三十

除夕

New Year's Eve

斗转星移世

Vicissitude

对于整个世界而言，15—16 世纪是变革的两百年。15 世纪之前，世界上的大部分财富集中在亚洲，权力中心也集中在中国、印度、波斯等老牌文明古国；15 世纪开始，被当时的古罗马帝国视为弱小、偏远、荒蛮的不列颠群岛及其他西欧地区，在结束了黑暗的中世纪、迎来文艺复兴之后，现代科学和资本主义开始兴起，成为现代军事、政治、经济、文化发展的摇篮。

15 世纪上半叶，西欧诸国怀着对富庶神秘的东方的向往，怀着探索与征服的目的启动了大航海时代。随着新大陆的发现，世界格局发生了剧烈变化，这一变化投射在全球贸易网上，是新航路的开辟以及殖民主义的登台。沿着“海上丝绸之路”而来的商船主人已经发生改变，但古老的中国却浑然不觉。由于明初实行的“海禁”政策，中国实际上已主动让出了“海上丝绸之路”的主导权。

For the whole world, the 15th and 16th centuries were two hundred years of innovation. Before the 15th century, most of the world's fortune was concentrated in Asia, and the power centers were also concentrated in China, India, Persia and other ancient civilizations. From the beginning of the 15th century, when the dark Middle Ages ended and came the Renaissance, the British Isles and other Western European regions, once regarded as weak, remote and savage by the ancient Roman Empire, had become the cradle of modern military, political, economic and cultural development, and the modern science and capitalism began to thrive.

In the first half of the 15th century, Western European countries started the era of great navigation with the longing for the rich and mysterious East and the purpose of exploration and conquest. With the discovery of the new continent, the pattern of world had changed dramatically. This change was projected on the

February / 2021.2

12

星期五

Friday

辛丑牛年 庚寅月

正月初一

春节

Spring Festival

global trade network, the initiation of new routes and the rise of colonialism. The roles of the owners of the merchant ships on the "Maritime Silk Route" had changed, but the ancient China did not realize it. In fact, the “ban on the maritime affairs” implemented in the early Ming dynasty made ancient China lost the dominant power of the "Maritime Silk Route".

19世纪通草水彩远眺港湾风景图。

Port scenery, watercolour on pith paper, 19th century.

清晚期布本琶洲塔珠江风景图。

Pazhou Pagoda beside pearl river, oil on canvas, Qing dynasty.

February / 2021.2

13

星期六

Saturday

辛丑牛年 庚寅月

正月初二

19世纪通草水彩船舶图。

Ships, watercolour on pith paper, 19th century.

February / 2021.2

14

星期日

Sunday

辛丑牛年 庚寅月

正月初三

片板难入海

No Ship Was Allowed to Sail

在欧洲人努力探索世界地图空白区域的时候，中国却在 14 世纪中叶开始实施严厉的“海禁”政策，导致海上贸易大幅萎缩。

明初禁海令的产生背景较为复杂，概括而言，一是因为明朝统治者认为农业是国之根本，故而重农抑商；二是朱元璋与张士诚、方国珍争天下的时候，江浙沿海一带百姓及泉州外国商团曾支持后者，朱元璋立国后，张、方余部逃至海上，不时骚扰东部沿海地区，让明政府颇为忌惮；三是当时东南海疆日本海盗（倭寇）十分猖獗，但明王朝主要精力放在消灭北方的元朝残余势力，无暇顾及海防；第四，海禁是朱元璋推行的一系列强化中央集权与君主专制政策的一环。几个因素结合在一起，使得朱元璋在 1368 年甫一建立明朝，便立刻实施海禁政策，并多次颁布诏令，强调“片板不许下海”。

While Europeans were trying to explore the uncharted areas of the world, China began to implement a strict policy which banned the maritime affairs in the middle of the 14th century, resulting in a sharp decline in maritime trade.

The background of the ban on the maritime affairs in the early Ming dynasty was complex. Generally speaking, the first reason was that the rulers of the Ming dynasty thought that agriculture was the foundation of the country, so they emphasized agriculture and restrained commerce. The second reason was that when Zhu Yuanzhang fought against Zhang Shicheng and Fang Guozhen for the sovereignty, the people in the coastal areas of Jiangsu and Zhejiang provinces and the foreign business groups in Quanzhou supported the latter. After Zhu Yuanzhang established the regime, Zhang and Fang fled to the sea and harassed the eastern coastal areas from time to time, worrying the Ming government; the third reason ☞

February / 2021.2

15

星期一

Monday

辛丑牛年 庚寅月

正月初四

was that Japanese pirates were rampant on South and East China Sea, but the Ming court mainly focused on eliminating the remnants of the Yuan dynasty in the north territory of ming court, and thus ignored the maritime defense; the fourth reason was that the ban was one of the steps of strengthening the centralization and autocratic monarchy policy implemented by Zhu Yuanzhang. In 1368, considering all these elements, as soon as the Ming dynasty was founded, Zhu Yuanzhang immediately carried out the policy to ban the maritime trade by issuing many imperial edicts, emphasizing that "not a single junk was allowed to sail on the sea".

澳门南湾。澳门是当时外国商船来华贸易的第一站。(图片采自《珠江风貌》)

South bay in Macao. Macao was the first stop through which the foreign merchant ships came to trade in China. (The picture is taken from *Views of the Pearl River Delta, Macao, Canton, and Hong Kong*.)

February / 2021.2

16

星期二

Tuesday

辛丑牛年 庚寅月

正月初五

广州迎宾地

Canton, the Place of Receiving Guests

私人商船在“海上丝绸之路”渐趋绝迹，官方主导的朝贡贸易随之兴起。所谓朝贡，是指古代海外诸国对中国统治王朝的纳贡。朱元璋将贸易系统和朝贡制度结合在一起，创制了朝贡贸易——明代唯一合法的对外贸易形式。明成祖朱棣尤其青睐朝贡贸易，不仅下令在广州、泉州（后移至福州）、宁波设置市舶司以方便贡使附带的货物交易，还在三地分别设置了怀远驿、来远驿和安远驿接待各国贡使及其随员。

明政府规定，广州接待占城、暹罗、满剌加、真腊等东南亚国家及西洋诸国贡使，泉州接待琉球群岛贡使，宁波接待日本贡使。广州是指定贡舶靠岸数最多的港口，因而怀远驿的建筑规模十分宏大，有房舍 120 间。到了嘉靖年间，由于倭患严重，福建、浙江的市舶司都被撤销，仅保留了广州市舶司。

Private merchant ships gradually disappeared on the "Maritime Silk Route", and the officially encouraged tribute trade thrived. The so-called “tribute” referred to the tribute paid to the Chinese ruling court by ancient overseas countries. Zhu Yuanzhang combined the trade system with the tribute system and created the tribute trade, the only legal form of foreign trade in the Ming dynasty. Zhu Di, the Third Emperor of the Ming dynasty, especially fond of the tribute trade, not only ordered the establishment of a municipal shipping administration department in Canton, Quanzhou (later moved to Fuzhou) and Ningbo to facilitate the trade of goods brought by the tribute envoys, but also respectively set up Huaiyuan post, Laiyuan post and Anyuan post to receive the tribute envoys and their entourages from all countries.

According to the Ming government, Canton received the envoys from Southeast ☞

February / 2021.2

17

星期三

Wednesday

辛丑牛年 庚寅月

正月初六

Asian countries such as Zhancheng, Siam, Malaga, Zhenla, and other Western countries; Quanzhou received the envoys from Ryukyu Islands and Ningbo received those from Japan. Canton was the designated port where most of the tribute ships docked, so the construction scale of Huaiyuan post consisting of 120 houses was very large. During the Emperor Jiajing's reign, due to the serious threat imposed by the Japanese pirates, municipal shipping administration departments of Fujian and Zhejiang were removed, and Canton's municipal shipping administration department was the only one to be retained.

Rain Water

雨水

February / 2021.2

18

星期四

Thursday

辛丑牛年 庚寅月

正月初七

18世纪《瓷器制运图》插画。图中描绘了清代粤商到景德镇拜会当地牙行瓷商（门上牌匾书有“然惠行”字样），采办外销瓷器的情景。

The Production and Transportation of Porcelain, 18th century. The illustration depicts the Cantonese merchants of the Qing dynasty came to Jingde town to visit the local porcelain merchants, and bought porcelain for export. (The name of the merchant house “Ranhui Hong” was written on the board under the roof.)

怀远开先声

The Emergence of the Hongs from Huaiyuan Post

怀远驿位于今广州荔湾区十八甫路东段，建于明成祖时期，不仅接待前来朝贡的外国使者及随员，同时也是货物交易场所。按惯例，历代前来朝贡的使者都会有商船随行，运载贡品及大宗贸易商品。由于明代禁止民间私自对外贸易，所以这些货物便由官府出面处理。最初，采取官府以高于实际货值的赏赐方式进行购买，称为“给赉”；后来又改为“抽买”，即官府先从中挑选质优者进行采购，剩下的允许与民间私人交易，但交易必须由官方组织，临时“招商发卖”，而且交易必须在怀远驿完成。久之，广州市舶司便设立了专办此事的牙行，而牙行负责办理贸易事务的牙人作为官方、外商与民间的中间人，已经隐隐具备“官商”的特征，这也正是后来十三行行商的特征之一。到明末海禁废弛的时候，广州已有卅六行承包外贸，为清代十三行的出现发了先声。

Huaiyuan Post was located in the eastern section of Shibafu Road, Liwan District, Canton. Built in the period of the Emperor Mingchengzu of the Ming dynasty, the post not only received foreign envoys and entourages who came to pay tribute, but was also used as a place for trade. According to the tradition, the envoys that came to pay tribute would be accompanied by merchant ships to carry tribute goods and bulk stock commodities. Because private trade was forbidden in the Ming dynasty, the goods were handled by the government. At first, the government rewarded the envoys with higher prices than the actual value of goods, which was called "granting". Later on, the government changed their way and only “bought the selected goods”, indicating that the government first selected the high-quality ones to be purchased, and the rest were allowed to be privately traded with residents, but the trade must be organized by the officials, who temporarily

February / 2021.2

19

星期五

Friday

辛丑牛年 庚寅月

正月初八

"encouraged the merchants to sell", and the trade must be completed in Huaiyuan post. For a long time, Canton municipal shipping administration department set up "Hongs" (merchant houses) to deal with this matter. As intermediaries between the government, foreign traders and the local people, the Hong workers, in charge of trade affairs, gradually bore the characteristic of "official merchants", which would become one of the characteristics of the Hong merchants of the Thirteen Hongs in the later period. By the end of the Ming dynasty, when the ban on the maritime affairs was lifted, there had already been 36 foreign trading categories in Canton, setting the first stage for the emergence of the Thirteen Hongs in the Qing dynasty.

19世纪广绣百鸟图挂饰。

Canton embroidery tapestry with patterns of birds, 19th century.

February / 2021.2

20

星期六

Saturday

辛丑牛年 庚寅月

正月初九

清末广绣寿眉鸳鸯图。

Canton embroidery with patterns of love birds, late Qing dynasty.

February / 2021.2

21

星期日

Sunday

辛丑牛年 庚寅月

正月初十

平台开粤府

Re-Opening of the Maritime Trade

明末渐渐松弛的海禁，到了清代又再次严格起来。清初的情况与明初颇为类似，西南的南明势力尚未平定，海上又有郑成功势力威胁。1654 年，清政府和郑成功和谈失败，眼看东线沿海战事将起，顺治帝便于 1655 年 6 月下诏："无许片帆入海，违者立置重典"；更于 1661 年进一步强令江、浙、闽、粤、鲁沿海居民内迁 30 至 50 里，设界防守，同时严厉禁止商民船只私自出海贸易，一旦被发现，犯事者处死，保甲连坐，地方官革职。

在先后平定"三藩之乱"及台湾后，国内局势稳定下来。1684 年，康熙开始全面放开海禁，包括官方设立的粤（广州）、闽（福州）、浙（宁波）、江（上海）四个海关分领对外贸易事务在内，全国实际与海外通商的口岸超过 100 个，沿着内河，西洋商船开进了内陆城市。

The ban on the maritime trade, which was gradually lifted at the end of the Ming dynasty, once again became severe in the Qing dynasty. The situation in the early Qing dynasty was quite similar to that of the early Ming dynasty. The forces of the Southern Ming dynasty in the Southwest have not yet been pacified, while there was threat from Zheng Chenggong's forces on the sea. In 1654, the peace talks between the Qing government and Zheng Chenggong failed. Considering the war on the east coast, Emperor Shunzhi issued an edict in June 1655: "not a junk is allowed to sail on the sea, and those who disobey it will be seriously punished"; in 1661, he further ordered the coastal residents of Jiangsu, Zhejiang, Fujian, Guangdong and Shandong provinces to move from 30 to 50 miles inwardly, set up a border for defense, and strictly prohibited the commercial and civilian ships from going out to the sea without permission. Once found, the offenders would be executed, while the ☞

February / 2021.2

22

星期一

Monday

辛丑牛年 庚寅月

正月十一

responsible guarantors put in jail and the local officials dismissed.

After pacifying the "rebellion of three feudal tribes" and the chaos in Taiwan, the domestic situation had stabilized. In 1684, the Emperor Kangxi began to thoroughly withdraw the ban on the maritime trade. In addition to the four official customs, Guangdong (Canton), Fujian (Fuzhou), Zhejiang (Ningbo) and Jiangsu (Shanghai), which were set up to handle foreign trade affairs separately, there were more than 100 ports actually traded with overseas countries. Through the inland rivers, the Western merchant ships entered the inland cities.

绘于18世纪的澳门南湾。清朝初期，清政府在澳门专门设置了澳门总口和海关监督的关部行台，并派遣专人管理澳门口岸的贸易，由此正式形成了清政府对澳门口岸的贸易管理机构。（图片采自《珠江风貌》）

This view of Macao was drew in the 18th century. In the early Qing dynasty, the Qing government set up a special customs office and a supervision bureau of Canton Customs in Macao, and sent special personnel to manage the trade, which officially formed the Qing government's trade management institution on Macao port. (The picture is taken from *Views of the Pearl River Delta, Macao, Canton, and Hong Kong.*)

February / 2021.2

23

星期二

Tuesday

辛丑牛年 庚寅月

正月十二

广州新时速

A New Speed in Canton

1670 年，在郑氏家族“通洋裕国”对外贸易政策的招徕下，英国东印度公司的“万丹号”和“珍珠号”到达台湾，与郑氏家族签订贸易合约，撕开了与东亚贸易的口子。1676 年，英国在中国大陆建立了第一个商馆——厦门商馆，它也是东印度公司在中国的总商馆。

东印度公司在厦门的贸易活动持续了 20 年，他们发现厦门官员征税勒索太重、本地“皇商”垄断严重。康熙三十七年（1698 年），为了扩大对外贸易，清政府“减粤海关额税”。第二年，东印度公司即派遣“麦士里菲尔德号”前往澳门，自海关监督处获得了到广州自由贸易的许可证，从此开启了中英贸易史上的“广州时代”。1704 年，东印度公司在中国的贸易中心转到广州，直到 1840 年第一次鸦片战争爆发。136 年间英国在中国的贸易史，实际就是英国广州商馆的历史。

In 1670, under the solicitation of the Zheng family's foreign trade policy which aimed at “getting along with the West to make the country wealthy”, the ships of the English East India Company, named "Pink Bantam" and "Sloop Pearl", arrived in Taiwan, signed a trade contract with the Zheng family, and began to trade with East Asia. In 1676, Britain established the first merchant house——the Amoy merchant house in mainland China, which was also the head quarter of the merchant houses of English East India Company in China.

The English East India Company's trade activities in Amoy lasted for 20 years. They found that Amoy officials' tax extortion was too heavy and there was a serious monopolization from the local “emperor's merchants”. In the thirty-seventh year of the Emperor Kangxi's reign (1698), in order to expand the foreign trade, the Qing ☞

February / 2021.2

24

星期三

Wednesday

辛丑牛年 庚寅月

正月十三

government “reduced the amount of tax of Canton Customs”. In the next year, the East India Company sent the ship "Macclesfield" to Macao, and obtained the license of free trade in Canton from the customs supervision office, creating the “Speed of Canton” in the history of Sino-British trade. In 1704, the English East India Company's trade center in China was transferred to Canton. Until the first Opium War in 1840, the history of British trade with China, lasting 136 years, had actually been the history of the British merchant houses in Canton.

清代广州十三行的茶叶店。茶叶贸易，是东印度公司在广东的主要大宗贸易之一。（图片采自《珠江风貌》）

This is a tea shop in the trading region. Tea trade was one of the major bulk stocks of the English East India Company in Guangdong province. (The picture is taken from *Views of the Pearl River Delta, Macao, Canton, and Hong Kong.*)

February / 2021.2

25

星期四

Thursday

辛丑牛年 庚寅月

正月十四

通商唯一口

One-Port Trade

1755年，为了扩大贸易范围，东印度公司派“中国通”洪仁辉带领商船到定海、宁波一带茶丝产地进行收购。他们发现浙海关的关税要比粤海关少，于是此后两年便绕开广州，直接前往宁波做生意，导致粤海关关税收入锐减。两广总督杨应琚担心贸易重心北移，便上奏朝廷请求调高浙海关关税，其中提到外国武装商船进入内陆数量太多、停留时间太长，容易与当地奸人勾结渔利，时日一久恐怕宁波、定海会成为另一个澳门。

这番话戳中了乾隆的心病，于是他下令提高浙海关的关税，但并没有禁止英商船前往宁波。杨应琚又上奏说，查明广州二十六家洋行办理外商贸易事务无不尽心尽力，洋商图利才跑到宁波去，洋船不至广东民生不举。乾隆最后决定，英国商船只许在广州上岸贸易，再不许进入浙江海域，此为“一口通商”的开始。

In 1755, in order to expand the trade scope, the English East India Company sent James Flint, “an expert on China”, to lead the merchant ships to the Dinghai and Ningbo, the production area of tea and silk, to purchase goods. They found that the tariff of Zhejiang Customs was less than that of Canton Customs, so they by-passed Canton and went directly to Ningbo to trade in the next two years, resulting in a sharp decline in tariff revenue of Canton Customs. Yang Yingju, the governor of Guangdong and Guangxi provinces, was worried that the trade center might be moved to the North, so he asked the court to raise the tariff of Zhejiang. He mentioned that there were too many foreign armed merchant ships berthing at the inland coast, and if they stayed there for too long, they might easily collude with the local profiteers and in the long run. Ningbo and Dinghai might become another Macao.

February / 2021.2

26

星期五

Friday

辛丑牛年 庚寅月

正月十五

元宵节

Lantern Festival

This remark hit the heart of the Emperor Qianlong, so he ordered to raise the tariff of Zhejiang customs, but did not forbid the British ships to go to Ningbo. Yang Yingju went on to say that his investigation showed that 26 merchant houses in Canton did their best to handle foreign trade affairs, so the foreign traders went to Ningbo was only to seek profits, and if the foreign ships did not stop at Canton, the local people might find it hard to make a living. Emperor Qianlong finally decided that the British merchant ships should only go ashore in Canton for trade, and were no longer allowed to enter the waters of Zhejiang. Thus it was the beginning of the "one-port trade".

February / 2021.2

27

星期六

Saturday

辛丑牛年 庚寅月

正月十六

清代木雕金漆“满大人”像。

Golden wooden Mandarin figure, Qing dynasty.

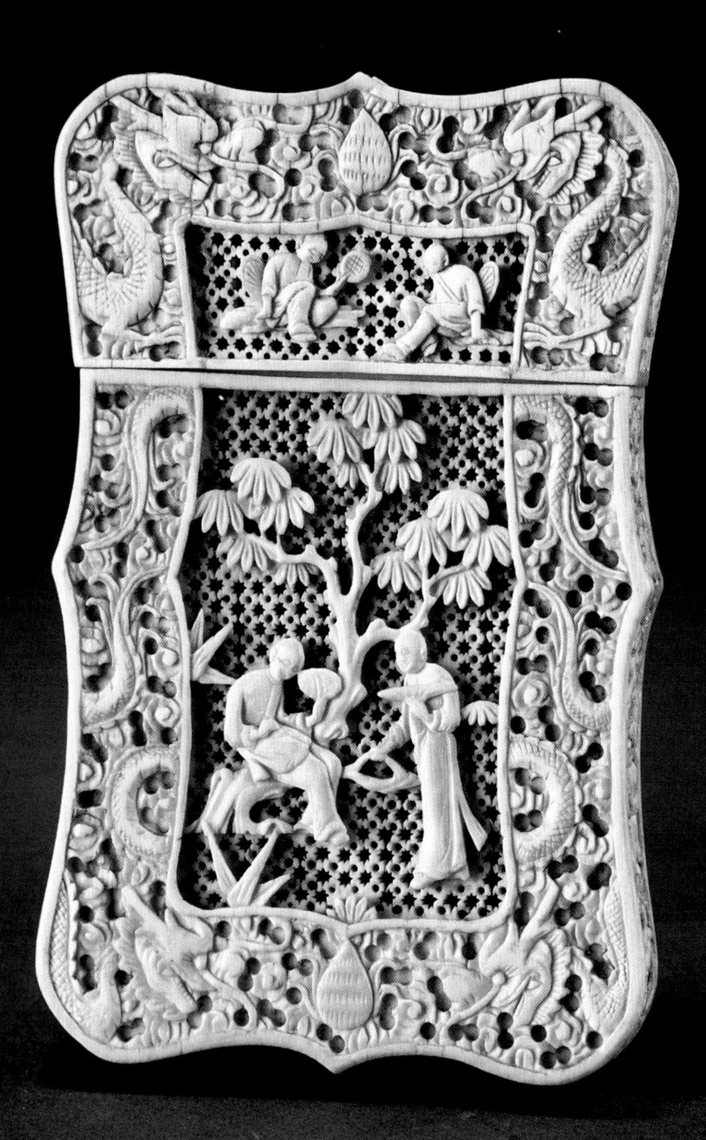

February / 2021.2

28

星期日

Sunday

辛丑牛年 庚寅月

正月十七

清末象牙雕开光镂空人物纹名片盒。

Ivory business card box with patterns of figures, late Qing dynasty.

March

一	二	三	四	五	六	日
1 十八	2 十九	3 二十	4 廿一	5 惊蛰	6 廿三	7 廿四
8 国际妇女节	9 廿六	10 廿七	11 廿八	12 植树节	13 二月	14 龙头节
15 初三	16 初四	17 初五	18 初六	19 初七	20 春分	21 初九
22 初十	23 十一	24 十二	25 十三	26 十四	27 十五	28 十六
29 十七	30 十八	31 十九				

告状到御前

Complaint before the Emperor

被限制了贸易自由的英国感到不忿，于乾隆二十四年（1759 年）派洪仁辉再去宁波尝试贸易，如果尝试失败就直接去天津，想办法面见乾隆皇帝告御状。洪仁辉来到浙江海域，果然被清朝水师拦截。他装作返回广州，实际转头北上去到天津，通过直隶总督将一纸诉状呈到乾隆面前，状告粤海关监督李永标贪污及刁难洋商，并代表东印度公司希望清政府改变外贸制度，实行多口通商。

乾隆相当恼火，一方面派人到广州彻查此事，革去李永标官职；另一方面认为洪仁辉冒犯了“天朝”威严，治了他一个勾结内地奸民、不遵守大清通商条例的罪名，判他在澳门圈禁三年，刑满驱逐回国；帮助洪仁辉写御状的中国人刘亚匾，则被砍头示众。

“洪仁辉事件”让清政府对于西方人的防御心理更加严重，很快便颁布了“防夷五事”，同时更加严格地实施 1757 年颁布的禁令，从此西洋商人只能在广州贸易，“一口通商”正式登上历史舞台。

Unsatisfied with the restriction on free trade, James Flint tried to go to Ningbo to trade again in the twenty-fourth year of the Emperor Qianlong's reign (1759). If the attempt failed, he would go to Tianjin directly and find a way to meet the Emperor Qianlong and complaint personally. James Flint came to the waters of Zhejiang province and was intercepted by the navy of the Qing court. He pretended to return to Canton but actually turned northwards to Tianjin. Through the general governor of direct supervision, he presented a petition to Qianlong, accusing a Canton Customs officer, Li Yongbiao of his corruption and creating difficulties for foreign traders. On behalf of the English East India Company, he hoped that the Qing government would change its foreign trade system and open more ports for foreign ☞

March / 2021.3

1

星期一

Monday

辛丑牛年 庚寅月

正月十八

trade.

On the one hand, the exasperated Emperor Qianlong sent officials to Canton to investigate the matter and dismissed Li Yongbiao; on the other hand, he thought that James Flint had offended the authority of the "Heaven Dynasty", and he was punished for colluding with the mainland's traitors and failing to follow the commercial regulations of the Qing court. He was sentenced to three years' imprisonment in Macao and would be deported to his country after serving his sentence. Liu Yabian, a Chinese who helped James Flint wrote the complaint, was beheaded and shown to the public.

The "James Flint affair" made the Qing government more defensive psychologically toward the Westerners. Soon, the Qing government issued the "five precautions on foreigners". At the same time, it reenforced the ban of 1757 more severely. Since then, the Western merchants could only trade in Canton, and the "one-port trade" officially appeared on the historical stage.

March / 2021.3

2

星期二

Tuesday

辛丑牛年 庚寅月

正月十九

云集在珠江口的外国商船。海上对欧美“一口通商”正式登上历史舞台后，西洋商人只能在广州进行贸易。（图片采自《珠江风貌》）

Foreign merchant ships on the estuary of the Pearl River. After the “one-port trade” came onto the historical stage, the Western merchants could only trade in Canton. (The picture is taken from *Views of the Pearl River Delta, Macao, Canton, and Hong Kong.*)

防夷重五事

Five Precautions on Foreigners

“防夷五事”全称《防范外夷规条》，共有五条，是清政府全面管制外商的第一个章程，主要内容为：一、禁止外国商人在广州过冬；二、外国商人到广州，应令寓居洋行，由行商负责稽查管束；三、禁止中国人借外商资本及受雇于外商；四、割除外商雇人传递信息之弊；五、外国商船进泊黄埔，酌拨营员弹压稽查。

“防夷五事”对外国商人的种种限制和防范可谓前所未有，不但禁止外国人与普通中国百姓接触，甚至也不许他们接触中国官吏，一切纳税、向政府呈递禀书、贸易等事宜都由行商代办。行商还负责安排外商在广州的居住，并对他们的活动进行监督，承担了相当部分的政府职责；但凡外商在中国境内有任何问题，政府便唯行商是问。行商这种“半官半商”的身份，正是清政府“以官制商，以商制夷”政策淋漓尽致的体现。

The full name of the “five precautions on foreigners” is “Regulations on Prevention of Foreigners”, which consisted of five articles in total. It was the first regulation announced by the Qing government to comprehensively take control of the foreign traders. The main contents were as follows: first, foreign traders were not allowed to stay in Canton during winter; second, foreign traders who came to Canton should live in the Hongs, namely the merchant houses, and were under the supervision and management of the Hong merchants; third, Chinese people were forbidden to borrow capital from foreigners or to be employed by them; fourth, the foreigners were not allowed to get their information conveyed; fifth, when foreign merchant ships berthed at Whampoa, the officers would be assigned to carry out inspections.

The restrictions and precautions imposed upon foreign traders by the “five ☞

March / 2021.3

3

星期三

Wednesday

辛丑牛年 庚寅月

正月二十

precautions on foreigners" were unprecedented. The foreigners were neither allowed to contact with ordinary Chinese people, nor with Chinese officials. All the matters such as paying taxes, submitting reports to the government, and trade must be handled by the Hong merchants, who were also responsible for arranging the residence for foreigners in Canton, supervising their activities, and assuming a considerable part of the government's responsibilities; whenever the foreign merchants caused any trouble in China, the Hong merchants would be scolded by the government. The identity of "half-official and half-merchant" was the most conspicuous embodiment of the Qing government's policy which aimed at "using the officials to control the merchants, and using the merchants to control the foreigners".

清代外销画中的粤海关。（图片采自《东西汇流》）

Canton Customs in the export paintings of the Qing dynasty. (The picture is taken from *East meets West*.)

March / 2021.3

4

星期四

Thursday

辛丑牛年 庚寅月

正月廿一

承袭牙行制

The Canton System Inheriting the Ya Hongs

牙行指代客商买卖从中说和的中介机构，出现于元代。明代牙行分为官牙和私牙两种，官牙由政府经营，私牙则是具有一定资金的商人，经政府批准，发给牙帖后从事买卖经纪活动。《大明律》规定，牙行无论官私，日常担负的责任有几种：向官府报告客商情况及货物数目；代客商买进卖出；代客商雇请车船人丁；代存货物；为客商提供食宿。

随着“隆庆开海”解除海禁，来华贸易的外国商船日益增多，广州等沿海地区的官牙分化出专门处理华夷贸易的市舶司官牙，并增加了代政府控制市场、管理商业的作用。嘉靖年间，随着广州市舶司互市交易场所迁至澳门，又出现身份类似于牙人的“揽头”。揽头来往于广州和澳门之间，外商违法，揽头须负连带责任被治罪。至此，十三行行商职责已基本形成。

“Ya Hongs”, the intermediary organizations which traded on behalf of their clients, appeared in the Yuan dynasty. In the Ming dynasty, these intermediaries were divided into two types: the official and private. The official ones were operated by the government, while the private ones were merchants who, with a certain amount of money, had been approved by the government and given the permission to take part in commercial activities as brokers. According to the law of the Ming dynasty, being official or private, the daily responsibilities of the “Ya Hongs” were as follows: to inform the government about the situation of the merchants and the number of goods; to buy and sell on behalf of the merchants; to hire the drivers and shipmen and store the goods for the merchants; to provide the merchants with foods and accommodations.

As the Emperor Qinglong lifted the ban on the maritime trade, the number of

March / 2021.3

5

星期五

Friday

辛丑牛年 辛卯月

正月廿二

foreign merchant ships coming to China for trade was increasing day by day. The official "Ya Hongs" in Canton and other coastal areas had been specialized into the official intermediaries of the municipal shipping administration department, dealing with the trade between China and the West. Its role of controlling the market and managing the trade on behalf of the government was reinforced. During the reign of the Emperor Jiajing, the trading area, Canton's municipal shipping administration department was moved to Macao, and there appeared other intermediaries called "Lantou", similar to those of "Ya Hongs". If a foreign trader violated the law, the intermediary should take the responsibility and be punished. By then, the responsibilities of the Hong merchants had been basically formed.

清末黑漆描金彩绘人物故事图折扇。二十一档黑漆描金扇骨，两大骨绘描金福庆有余纹饰；扇面顶部彩绘瑞兽珍禽、八宝、鱼蟹蔬果、花鸟蝴蝶；中部主体彩绘通景人物故事图；底部饰描金花卉、卷草纹。此扇绘画色调鲜艳，繁而不乱，黑漆描金彩绘工艺相结合较少见。

Black lacquer folding fan with gold-painted polychrome figures was made in late Qing dynasty. There are 21 sticks painted black with golden patterns, and on two of the sticks there are depictions of auspicious patterns; the top of the fan is painted with colourful auspicious animals, rare birds, treasures, fish, crabs, fruits, vegetables, flowers, birds and butterflies; the main part in the middle is painted with a story scene where the characters are in the garden; the bottom of the fan is decorated with golden flowers and rolling grasses. The colour of the painting is bright, complex but not disorderly, and the combination of golden patterns on black lacquer with polychrome painting is rarely seen.

March / 2021.3

6

星期六

Saturday

辛丑牛年 辛卯月

正月廿三

7

星期日

Sunday

辛丑牛年 辛卯月

正月廿四

19 世纪黑漆描金扇骨广绣花鸟纹圆形执扇。

Black lacquer fan with gold-painted patterns of flowers and birds on Canton embroidery, 19th century.

专行洋货事

The Hongs of Foreign Goods

康熙二十三年（1684 年）起，清政府放开海禁，并设置四个海关取代市舶司。在通商管理制度方面，则沿袭明代前例，用牙行商人主持经营对外贸易。《粤海关志》记述："设关之初，番舶入市者，仅二十余柁，至则劳以牛酒，令牙行主之，沿明之习，命曰十三行。"需要强调的一点是，十三行并不是设置海关之初就出现的，而是稍晚一点（1686 年）。由于广州在明代就是朝贡贸易最活跃的地区之一，因而开关之初，大量国内外的货物都运到广州。但与之相对的是从事牙行的人并不多，导致货物一时不能报关缴税，堆积如山。为了适应开关后对外贸易需求，保证关税征收，清政府以法令形式发布《分别住行货税》，把从事国内沿海贸易牙人和从事对外进出口贸易牙人的活动范围及性质划分开来，设立金丝行、洋货行。

Since the twenty-third year of the Emperor Kangxi's reign (1684), the Qing government had lifted the ban on the maritime trade and set up four customs to replace the municipal shipping administration departments. In terms of commercial management system, it followed the example of the Ming dynasty, using the Ya Hong merchants to manage foreign trade. *The Records of the Canton Customs* mentions: "at the beginning of the establishment of the customs, there were only 20 ships that came to trade. When the ships arrived, the Ya Hongs received them and served them with wine. This is to follow the practice of the Ming dynasty, and the Hongs were called the Thirteen Hongs." It should be emphasized that the Thirteen Hongs did not appear at the beginning of the establishment of customs, but came into being later (1686). Since Canton was one of the most active areas of tribute trade in the Ming dynasty, many domestic and foreign goods were transported to Canton at the

March / 2021.3

8

星期一

Monday

辛丑牛年 辛卯月

正月廿五

国际妇女节

International Women's Day

opening of the customs. However, contrasted to the large amount of goods, there were not many Ya Hong intermediaries, so that the goods could not be declared and the tax could not be calculated, and the goods were piled up like a mountain. In order to meet the needs of foreign trade after the opening of the customs and ensure the collection of customs duties, the Qing government issued the Separation of the Domestic Tax and Foreign Trade Tax in the form of decrees, which, according to their activity scopes and characteristics, separated the domestic intermediaries who traded along the coastal area from those who engaged in foreign import and export trade. Thus the "Hongs of Gold Silk" and and the "Hongs of Foreign Goods" were established.

十三行商馆区内最繁华的街道之一：同文街。同文街位于丹麦馆及西班牙馆之间，街道两旁均为中国人所开设的各类店铺。

The busiest street in the trading region was New China Street. The New China Street was between the Danish Hong and Spanish Hong; there were all kinds of shops run by the Chinese on both sides of the street.

March / 2021.3

9

星期二

Tuesday

辛丑牛年 辛卯月

正月廿六

“金丝”与“洋货”

“Gold Silk” and “Foreign Goods”

金丝行和洋货行是清代粤海关对从事国内贸易和对外贸易商行的划分。1686年，广东巡抚李士桢发布《分别住行货税》，对两者进行了区别划分，如“来广省本地兴贩，一切落地货物，分为住税报单，皆投金丝行，赴税货司纳税”“其外洋贩来货物及出海贸易货物，分为行税报单，皆投洋货行，俟出海时，洋商自赴关部（粤海关）纳税”。所谓“住”就是指国内贸易，“行”指对外贸易，所以十三行商人也被称为“行商”。

同明代的牙行一样，从事行商的人必须身家殷实，同时需要地方官府核准后发给行帖。倘若某人同时经办国内和国外两种贸易，也必须分别设行，各立招牌，不许混同办理。乾隆初年，“金丝行”改名“海南行”，“洋货行”改为“外洋行”（简称“洋行”），广东十三行行商制度又叫“广东洋行制度”。

The “Hongs of Gold Silk” and “Hongs of Foreign Goods” were the divisions between domestic and foreign trade merchant houses set by the Canton Customs in the Qing dynasty. In 1686, Li Shizhen, the governor of Guangdong province, issued the Separation of the Domestic Tax and Foreign Trade Tax, distinguishing the two divisions. For example, “if someone comes to Guangdong province to trade, all the goods landed on the ground should be declared in the category of domestic tax, which would be collected by the Hongs of Gold Silk and the trader should pay taxes in the tax bureau. If a foreign merchant comes or goes with import or export cargoes, then it belongs to the category of foreign trade tax, which should be collected by the Hongs of Foreign Goods. The trader should personally pay the tax in the Canton Customs before going out to the sea”. The so-called “Zhu” refers to domestic trade and “Hongs” refers to foreign trade, so the merchants of the Thirteen ☞

March / 2021.3

10

星期三

Wednesday

辛丑牛年 辛卯月

正月廿七

Hongs were also known as "Hong merchants".

In the Ming dynasty, people engaged in commercial activities must be wealthy, and at the same time, they needed to be approved by the local government and then issued with permission certificates. If someone dealt with domestic and foreign trade at the same time, they must set up separate merchant houses with different brands respectively. Different trades could not be conducted by the same merchant house. In the early years of Emperor Qianlong's reign, the name "Hongs of Gold Silk" was changed into "Hongs of Hainan", the name "Hongs of Foreign Goods" was changed into "Foreign Hongs" (whose abbreviation was "Hongs"), and the trade system of the Thirteen Hongs in Guangdong province was called "Canton system".

靖远街是十三行商馆区内另一条著名的商业街。这里商贾云集，外销扇和其他广作外销艺术品由此销往海外。

Jingyuan Street, also known as Old China Street, was a famous commercial street in the trading region, where there were a lot of merchants. Export fans and other artworks were exported overseas from here.

March / 2021.3

11

星期四

Thursday

辛丑牛年 辛卯月

正月廿八

藩商有红顶

The Merchants with Red-Top Official Hats

最早行商的构成“成分”还是比较复杂的，大致有以下几种：第一，由广州府及珠江三角洲一带原本充当海贸牙行的商人转充而来，隶属于广商集团；第二，活跃于广东沿海地区的徽州、泉州商人集团；第三，清初出现于广东地区的“藩商”。

所谓“藩商”，是在清初实施海禁和藩镇军事割据统治下产生于广东的、从事海上走私贸易的商人，他们背后的实际靠山正是平南王尚可喜之子尚之信。最有名的大“藩商”名叫沈上达，系藩王府参将、贸易总管。他在广东各级政府官员庇护下，大肆走私，为藩王府赚取了大量白银，也使自己成为显赫一时的富商。这群亦官亦商的“藩商”，凭借着背后的政治军事和封建经济特权，成为清代前期广东一个集政治和经济于一体的特权阶层，对皇权产生了很大威胁，这也是康熙决意平藩的一大原因。

The “composition” of the earliest Hong merchants were relatively complex with the following origins: the first type were the traders in Canton as well as around the Pearl River Delta, originally serving as the intermediaries of maritime trade, and they were affiliated to the Cantonese merchant groups; the second type, active in the coastal areas of Guangdong province, were from the merchant groups of Huizhou and Quanzhou; the third type were the “vassal merchants” who appeared in Guangdong area in the early Qing dynasty.

Under the rule of the military separatism of the vassal towns, the so-called “vassal merchants” appeared in Guangdong province in the early Qing dynasty when the maritime trade was banned. They were engaged in the smuggling trade on the sea. The actual backing behind them was Shang Zhixin, the son of Shang Kexi, the

March / 2021.3

12

星期五

Friday

辛丑牛年 辛卯月

正月廿九

植树节

Tree Planting Day

king who pacified the South. The most famous "vassal merchant" was Shen Shangda, who was the general counsellor and trade manager of the vassal government. Under the protection of the officials of all levels in Guangdong government, he engaged in smuggling and earned a large amount of silver for the vassal government and made himself a rich merchant for a time. This group of "vassal merchants", who were both officials and merchants, became a privileged class of political and economic integration in Guangdong in the early Qing dynasty by virtue of the political, military and feudal economic privileges behind them, posing a great threat to the imperial power, which was also a major reason for Emperor Kangxi's determination to suppress the vassals.

March / 2021.3

13

星期六

Saturday

辛丑牛年 辛卯月

二月初一

清末铜胎画珐琅佛教故事图葫芦瓶。

Enamel cucurbit-shaped vase with illustration of the Buddhist story, late Qing dynasty.

March / 2021.3

14

星期日

Sunday

辛丑牛年 辛卯月

二月初二

龙头节

Dragon Head-Raising Day

清代铜胎画珐琅缠枝花卉纹双耳狮钮盖三足薰炉。

Enamel incense burner with three-legged stand, and patterns of branches and flowers, Qing dynasty.

李士桢入粤

Li Shizhen, an Outstanding Official

尽管明代的广州在朝贡贸易和经济方面都取得了不错的成就，但明清更替之际，特别是顺治四年（1647 年）到康熙二十年（1681 年）三十余年间，在先后经历了南明小王朝联合张献忠余部与清军在岭南地区的拉锯战、明降将李成栋的降而又反、尚可喜和耿继茂两大藩王的平叛及割据统治、清初的禁海令后，人口锐减，社会经济遭到严重摧残，连康熙皇帝也不得不承认“广东民人，为王下兵丁扰害甚苦，失其生理”。

从民生凋敝到经济复苏，继而成为清前期中国唯一的对外贸易大港，时任广东巡抚的李士桢功不可没。正是他在任上采取的一系列整顿吏治、招抚商民、恤商裕课等措施，将广东地区的政治、民生和经济进行了全面修复，更创建了行商制度——广州十三行，对广东对外贸易的发展，意义尤为重大和深远。

Although Canton made good achievements in the tribute trade and economic development in the Ming dynasty, when the Ming regime was replaced by the Qing regime, especially during the thirty years from the fourth year of Emperor Shunzhi's reign (1647) to the twentieth year of Emperor Kangxi's reign (1681), it successively suffered from two long-lasting battles between Zhang Xianzhong's remaining force and the Qing army in Lingnan area, the surrender and rebellion of Li Chengdong, a general of the Ming court, as well as the revolt of the separatist regime of two vassal kings, Shang Kexi and Geng Jimao. Because of the ban on the maritime trade in the early Qing dynasty, the population sharply declined and the social economy had been destroyed. Even the Emperor Kangxi had to admit that "the Guangdong people suffered a lot under the soldiers of the kings and lost their livelihood".

Saving people's livelihood from desolation and recovering the economy, ☞

March / 2021.3

15

星期一

Monday

辛丑牛年 辛卯月

二月初三

Li Shizhen, the governor of Guangdong province at that time, made a great contribution, and Canton became the only port for foreign trade in China in the early period of the Qing dynasty. It was him that took a series of measures in his tenure, such as rectifying and administrating the officials, encouraging the traders, appeasing the people, and offering traders with subsidies and making clear the taxation. Thus the political condition, people's livelihood and economy of Guangdong province were comprehensively recovered and the Canton system——the Thirteen Hongs was established, shedding a great and far-reaching influence on the development of foreign trade in Guangdong.

十三行商馆区的美国花园和圣公会教堂。（图片采自《珠江风貌》）

The American garden and the Anglican church in the trading region. (The picture is taken from *Views of the Pearl River Delta, Macao, Canton, and Hong Kong*.)

March / 2021.3

16

星期二

Tuesday

辛丑牛年 辛卯月

二月初四

缔造十三行

Establishing the Thirteen Hongs

广东巡抚李士桢对于岭南地区最卓著的贡献是创建了十三行。重农轻商的思想在中国古代根深蒂固，而李士桢在当时可谓少有的具备开阔眼界的官员。据《抚粤政略》记载，李士桢到广东上任半年间，呈上的"奏疏"和发布的"符檄""文告"共54份，和商业有关的23份，足见其对商业的重视程度。

李士桢69岁辞官时，士民罢市，夹道送行，并为他立祠于五仙门外，又入祀省城名宦祠。李士桢后来定居北京通州，其子李煦深受康熙重用，被任命为苏州织造，并兼管盐务，与曹雪芹祖父曹寅同朝为官，更是曹雪芹祖母的娘家，李家也是曹雪芹笔下荣宁二府的原型之一。

Li Shizhen, the governor of Guangdong province, made the most outstanding contribution to the Lingnan region by creating the Thirteen Hongs. The thought of emphasizing agriculture while neglecting commerce was deeply rooted in ancient China, and Li Shizhen was an official with broad vision, which was rare at that time. According to *The Political Strategy to Appease Canton*, during his tenure of the first half year in Guangdong, Li Shizhen had presented 54 reports, official documents and statements, including 23 notices related to the commercial issues, showing his attention to commerce.

Li Shizhen resigned at the age of 69. At that time, the commercial officials and the businessmen all stopped working, and stood on both sides of the street to bid him farewell. A temple outside the Five Immortals Gate was built in commemoration of him, and he was worshiped in the Temple of Famous Officials in the provincial city. Li Shizhen later settled in Tongzhou, Beijing. Li Xu, his son, highly valued by the Emperor Kangxi, was appointed to be the administrator of the textile industry in

March / 2021.3

17

星期三

Wednesday

辛丑牛年 辛卯月

二月初五

Suzhou province and also to be in charge of the salt trade. He was the colleague of Cao Yin, Cao Xueqin's grandfather, who served as an official in the same period. He was also a member of the family of Cao Xueqin's grandmother. The Li family was also one of the prototypes of the Rong and Ning mansions in Cao Xueqin's book.

(Note: Cao Xueqin was the author of the famous book *The Dream of the Red Mansion.*)

在清代广州画家创作的外销画中，有大量描绘十三行的作品，从各个角度展现了当年万商云集的盛况。（图片采自《香江遗珍》）

The prosperous scene of the trading region where numerous merchants gathered can be seen in the export painting. (The picture is taken from *The Charter Legacy: A Selection of the Charter Collection*.)

March / 2021.3

18

星期四

Thursday

辛丑牛年 辛卯月

二月初六

行随江岸阔

The Wide Riverbank

在中国贸易史上，十三行代表着多重含义，它既是一处特定的办理对外贸易的商区，也是一种管理外商来华贸易的制度，还是一群专门从事对外贸易业务的商人公行组织。

十三行依江而建。由于珠江泥沙堆积作用，珠江北岸不断向南延伸，十三行商业区便也随之南移，依然建于江畔。今天的十三行旧址，已经远离珠江北岸，范围大致在东跨仁济路、西至杉木栏路、南临珠江边、北抵十三行路的广州文化公园一带。

作为清政府闭关锁国政策下唯一的对外贸易口岸，十三行商区在第一次鸦片战争之前极为繁华，拥有通往欧洲、拉美、南亚、东洋和大洋洲的环球贸易航线，年上缴税银超过百万。这些税银中，有一部分专门作为皇室的经费开支。因此在皇帝眼中，这个远离京城的商埠不啻于取之不尽的宝库，因而十三行也被誉为“金山珠海，天子南库”。

In the history of China's trade, the Thirteen Hongs represent multiple meanings, not only referring to a certain commercial region for foreign trade and a system for managing foreign trade in China, but also a merchant guild set up by a group of merchants who specialized in foreign trade.

The merchant houses were built along the riverbank. Due to the sediment of the Pearl River's mud and sand, the northern bank of the Pearl River continued to extend to the south, and the trading region was also moved to south, but still adjacent to the riverbank. Today's site of the trading region is far away from the northern bank of the Pearl River, ranging from Renji Road in the east, Shanmulan Road in the west, the Pearl River in the south, and Guangzhou Culture Park along

March / 2021.3

19

星期五

Friday

辛丑牛年 辛卯月

二月初七

the Thirteen Hongs Road in the north.

As the only foreign trade port under policy of isolationism of the Qing government, the trading region was very prosperous before the first Opium War, with global trade routes leading to Europe, Latin America, South Asia, East Asia and Oceania, and providing the court with more than one million tael of silver as tax income annually. Some of the tax paid by silver was specially used as the royal family's expenses. Therefore, in the eyes of the emperors, this commercial port far away from the capital was an inexhaustible treasure house. Therefore, the Thirteen Hongs were also known as the "mountain of gold, sea of pearls and the emperor's treasure house in the South".

19世纪通草水彩园景仕女图。

Maidens in the garden, watercolour on pith paper. 19th centruy.

19世纪纸本水彩园景仕女图。

Garden scene and maidens, watercolour on paper, 19th century.

春分

Spring Equinox

March / 2021.3

20

星期六

Saturday

辛丑牛年 辛卯月

二月初八

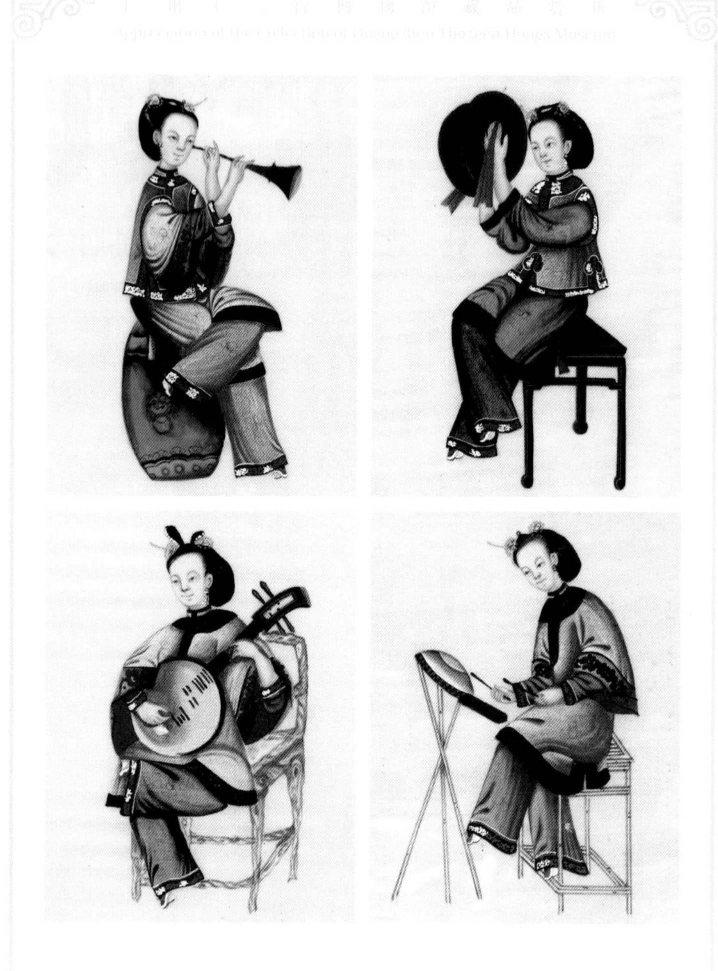

19世纪通草水彩仕女奏乐图。

Ladies playing the musical instruments, watercolour on pith paper, 19th century.

March / 2021.3

21

星期日

Sunday

辛丑牛年 辛卯月

二月初九

大防十三夷

Precaution on the Foreigners in the Trading Region

即便清政府开辟了十三行商区，但本质上始终抱持着“华夷大防”的心理。十三夷馆正是通商与华夷大防互相妥协下的产物。它是由政府指定地点、十三行行商出资建造的房屋建筑群，以出租的方式为来华贸易的外国商人提供住宿、储货和交易场所。

为了方便生意，加上十三行行商对外国人承担监督责任，十三夷馆皆建在行商行馆附近，毗邻当时的珠江，即今天广州十三行路以南、人民南路以西、珠江河以北的区域（今广州文化公园一带）。夷馆名“十三”则是以“十三行”命之，并不是指具体有十三座建筑。据《华事夷言》记载：“十三间夷馆……内住英吉利、弥利坚、佛兰西、领脉、绥林、荷兰、巴西、欧色特厘阿、俄罗斯、普鲁社、大吕宋、布路牙等之人，按此即所谓十三行也。”至今，我们仍旧能从传世的外销画中看到十三夷馆前飘扬的不同国家的国旗中，一窥当时盛况。

Even though the Qing government opened up the trading region, in fact, it always adhered to the mentality of “taking precaution on the the local people and foreigners”. The trading region, exactly as the result of the compromise between the trade and precaution, was built to be an architectural complex consisting of houses at a place designated by the government with the funds provided by the Hong merchants. Providing accommodations, warehouses and places for trade, it was rented by foreign traders who came to China.

To facilitate business, and for the Hong merchants to take the responsibility of supervising the foreigners, the merchant houses were all built near the building of the merchant guild, adjacent to the Pearl River at that time. The location covers from today's southern part of the Thirteen Hong Road, western part of Renminnan Road

March / 2021.3

22

星期一

Monday

辛丑牛年 辛卯月

二月初十

and northern area of the Pearl River (around Guangzhou Culture Park today). The "thirteen merchant houses" was known as the "Thirteen Hongs", but it did not mean that there were exactly thirteen buildings. *The Chinese Affairs and the Foreigners' Comments* mentions that "in the merchant houses ... there are People from England, America, France, Denmark, Sweden, the Netherlands, Brazil, Austria, Russia, Prussia, Spain, Portugal and so on. Thus, these were the merchant houses of the Thirteen Hongs". Up till now, we can still see the national flags of different countries in front of merchant houses in the export paintings which have been handed down from the previous time, and have a glimpse of the grand occasion at that time.

March / 2021.3

23

星期二

Tuesday

辛丑牛年 辛卯月

二月十一

远眺十三行商馆。1743 年，广东十三行第一次遭遇大火灾。广东顺德诗人罗天尺目睹了这场灾难，有感而发，写下 “雄如乌林赤壁夜鏖战，万道金光射波面”的诗句，可见当时火势之凶猛。（图片采自《珠江风貌》）

The trading region seen from a far perspective. In 1743, the trading region was damaged by a conflagration for the first time. Luo Tianchi, a poet from Shunde, Guangdong, witnessed the fire and was deeply touched, so he wrote a poem: "the conflagration is so ferocious that it bears the likeness of the fire attack of the fierce battle in the ebony forest, which made the red cliff tinged with the flaming blaze; tens of thousands of golden light beams are shooting onto the ripple." The poem describes the horrible scene of the huge fire at that time. (The picture is taken from *Views of the Pearl River Delta, Macao, Canton, and Hong Kong*.)

华筑有洋风

The Chinese Buildings with the Western Style

十三夷馆起初由行商兴建，所以建筑多为中国传统风格。后来在三次火灾后重建的过程中，外国商人从参与到主导夷馆设计建造，使建筑带上了显著的西洋风格。

美国人威廉·C. 亨特在他的《广州番鬼录》和《旧中国杂记》(1882年)里，对夷馆有详细的描述：每座夷馆的正面都朝着南边的珠江，各国商馆由西向东排列，西边以联兴街为界，东边以西濠分隔，中间则以靖远街、同文街和新豆栏街分割成三个区域。商馆背后是一条东西向的长街，即十三行街。十三行街北面，与同文街相望的一座宽大而漂亮的建筑是“公所”，也叫“洋行会馆”。这是行商公行用于接待客人和办事的地方，属于公产。外国商人倘若想办理贸易事务，或者了解清政府颁布的最新规定或税则，都必须要来这里。

At first, the merchant houses were built by the Hong merchants, so most of the buildings were built in traditional Chinese styles. Later on, because of three conflagrations, the merchant houses were rebuilt. Foreign traders participated and even played the leading role in the reconstruction, making the buildings take on a conspicuous Western style.

William C. Hunter, an American, described the merchant houses in detail in his books *Fankwae at Canton* and *Bits of Old China* (1882): the merchant houses representing different countries, were arranged from the west to the east, facing toward the southern shore of the Pearl River, and demarcated by Lianxing street in the west, and Xihao in the east, while the region was divided into three areas by Old China Street, New China Street and Hog Lane. Behind the merchant houses there was a long street running from the east to the west, namely, the Thirteen Hongs Street, to the north of which, a large and beautiful building overlooking New China

March / 2021.3

24

星期三

Wednesday

辛丑牛年 辛卯月

二月十二

Street was called "Consoo", also known as the "headquarter of the merchant guild", a public property for the merchant guild to receive their guests and deal with business affairs. If foreign traders wanted to deal with the trade affairs, or learn about the latest regulations or taxation issued by the Qing government, they must come here.

广州海幢寺于乾隆末年对外国人开放。(图片采自《大清帝国城市印象——19世纪英国铜版画》)

Haizhuang Temple in Canton was open to foreigners in the late period of Emperor Qianlong's reign. (The picture is taken from *Impressions of 19th Century Chinese Cities, Allom's Painting*.)

March / 2021.3

25

星期四

Thursday

辛丑牛年 辛卯月

二月十三

南洋传骑楼

Qilou, Buildings with Porticos

十三夷馆也被认为是如今岭南文化的标志之一——骑楼的早期雏形。由外国商人参与兴建的夷馆，为了适应广东潮湿炎热的天气，通常都会采用外廊式设计，有着向河边凸出的柱廊和露台，以便纳凉赏景。19 世纪末 20 世纪初，在十三夷馆被烧毁数十年后，第一批“下南洋”“上金山”的移民带着资金回乡创业经商。为了方便做生意，他们参考夷馆建筑的功能布局，再将盛行于南洋的建筑形式和岭南本土建筑结合起来，形成了一种宜商宜居、生意生活两兼顾的新的建筑形式——骑楼。

骑楼的内部沿用了夷馆的功能设置，二楼及以上作为住家，一楼则根据自身需要改为店铺；同时将相邻骑楼的外廊延伸打通，将外廊的属性从私人领域转变为供人行走的公共空间，仿佛一条带顶棚的长街，店铺就可以全天候营业，风雨无阻。

The merchant houses——the early prototypes of *Qilou* (a kind of buildings with porticos)——were also considered to be one of the symbols of Lingnan culture. In order to adapt to the humid and hot weather in Guangdong Province, the merchant houses jointly built by foreign traders usually adopted the outer porticos, with colonnades and terraces stretching toward the riverbank, so it was possible for the people to enjoy the coolness and scenery. At the end of the 19th century and the beginning of the 20th century, decades after the merchant houses were burned down, the first batch of migrants who “went down to the Southeast Asia” and “went up to San Francisco” returned home with funds to start businesses. To facilitate business, they adapted the functional layout of the merchant houses, and then combined the architectural form popular in Southeast Asia with the local buildings in

March / 2021.3

26

星期五

Friday

辛丑牛年 辛卯月

二月十四

Lingnan, creating a new architecture–*Qilou*, which was suitable for doing business and daily life.

The interior of *Qilou* followed the functional settings of the merchant houses. The second floor and the floors above were used by the household, while the first floor was changed into shops according to its own needs. At the same time, the outer corridors of the adjacent buildings were connected together, and thus the private corridors were changed into an outer portico which was like a long street with a roof above, forming a public space for people to walk, so that the shops could be open in all kinds of weather, regardless of rain or wind.

March / 2021.3

27

星期六

Saturday

辛丑牛年 辛卯月

二月十五

民国银累丝烧珐琅彩兰花纹盖罐。盖罐通体采用银累丝网状纹铺底，罐盖嵌珐琅彩兰花纹，配绿松石、红珊瑚 圆珠及银珠点缀其间；盖钮为花型圆球状装饰，顶饰红珊瑚；罐身镶嵌珐琅彩兰花、竹叶纹于椭圆形开光之内。这个盖罐做工精巧，银累丝工艺复杂，用料名贵，集中国传统工艺美和实用性于一身。

Silver filigree enamel covered can with floral patterns was made in the period of the Republic of China. The whole body of the can is laid with the silver filigree net pattern, and the cover is inlaid with enamel cymbidium patterns, and decorated with turquoise, red coral beads and silver beads; the cover knob is decorated with ball-shaped flowers, with the red coral on the top; the oval open windows on the can are inlaid with enamel cymbidium and bamboo-leaf patterns. This covered can is exquisite in workmanship and precious in materials with complex filigree techniques. It integrates the beauty and practicability of traditional Chinese craftsmanship.

March / 2021.3

28

星期日

Sunday

辛丑牛年 辛卯月

二月十六

清代铜胎银浮雕珐琅彩花卉纹方形盖罐。

Silver filigree pot with enamel patterns of orchids, Qing dynasty.

限夷生买办

Compradors

十三夷馆并非人人可住。根据规定，来广州的外国商船只有部分高级职员可以入住夷馆，妇女及其他船员只能在黄埔港附近指定的地点居留。由于洋流和季风的缘故，外国商船每年春夏到来，秋冬离开。为了避免外国人随意逗留，清政府规定外国人不许在广州过冬，如果确实有特殊情况，只准待在澳门，但次年必须离境。

即便入住夷馆，也不代表会受到优待。按规定，外国人不许进入广州城内（夷馆当时设在城墙之外的江边），只准在夷馆附近活动。到了乾隆末年，始准每月三次到隔海的陈家花园（后改往花地）和海幢寺两处游玩，但要洋行的通事随行约束。外国商人的所有贸易事务和对公事务由行商代理，所有日常生活受行商约束，渐渐的，一种名为“买办”的阶层诞生了。

Not everyone could stay in the merchant houses. According to the regulations, only some senior officers who came with foreign merchant ships arriving at Canton could stay inside, while women and other sailors could only stay in the designated places near Whampoa port. Because of the ocean current and monsoon, foreign merchant ships came here in spring and summer every year, and left in autumn and winter. To prevent foreigners from arbitrarily staying here, the Qing government stipulated that foreigners were not allowed to stay in Canton during winter. If there were any special circumstances, they could only stay in Macao, but must leave in the next year.

Even if they could stay in the merchant houses, it did not mean that they might be treated well. According to the regulations, foreigners were not allowed to go into the inner city of Canton (the merchant houses were located at the riverside outside

March / 2021.3

29

星期一

Monday

辛丑牛年 辛卯月

二月十七

the city wall), and their activities were constrained near the merchant houses. At the end of the Emperor Qianlong's reign, three monthly visits to the Chen Garden (later changed into Huadi) and Haizhuang Temple were reduced to two, with the supervision and accompany of the linguists from the merchant houses. All trade and official affairs were represented by the Hong merchants, and the foreigners' daily life was also supervised by the Hong merchants. Gradually, a class called "compradors" came into being.

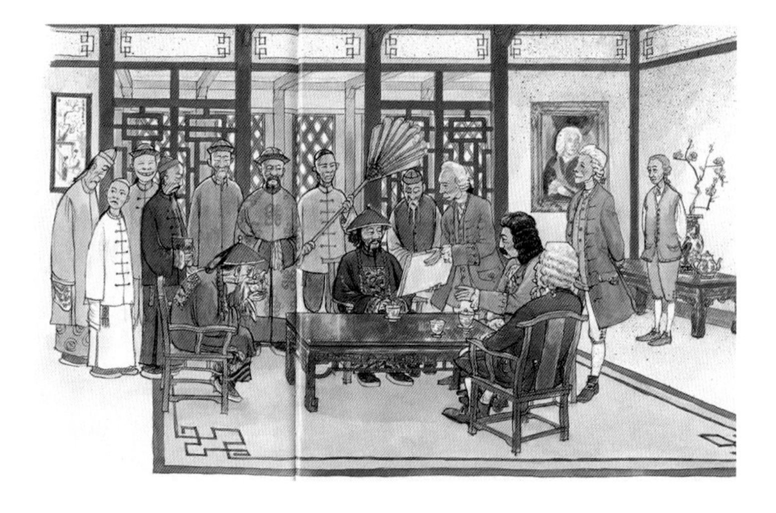

这幅绘画描绘了中外商人在夷馆洽谈生意时的情景。除了政府官员和外商，画中还出现了“买办”。（图片采自 *De långa resan*）

The painting shows that the Chinese and foreign merchants were having a business talk in a merchant house. Apart from the officials and foreign merchants, there were compradors in the painting. (The picture is taken from *De långa resan*.)

March / 2021.3

30

星期二

Tuesday

辛丑牛年 辛卯月

二月十八

买办万事通

Compradors Who Dealt with Everything

诞生于清代前中期行商制度之下的买办阶层，虽然有着服务外国商人的属性，但更多是受政府控制，然后又控制着外商，即所谓的“以官制商，以商制夷”。

十三行时期的买办大致分为两类：一类是专为停泊在黄埔、澳门水域的外商船只采买物料及食品的商船买办；一类是在外商商馆中代外商管理总务及现金的商馆买办。商馆买办由为外商服役的仆役头目演化而来，身份比普通仆役特殊，必须由行商、通事结保，并向粤海关领取牌照才能充当。在部分寓居夷馆的外国商人眼中，买办的重要性和权柄之大甚于行商，主要的一点就是夷馆中所有中国服务人员，包括外商会计、厨子、仆人、苦力等，都是买办的“自己人”。

Born in the business system of the early and middle Qing dynasty, and largely controlled by the government, the comprador class, served and meanwhile controlled the foreign traders. This was the so-called policy that “used the officials to control the merchants and used the merchants to control the foreigners”.

There were two types of compradors in the period of Canton trade: the first were the ship compradors who purchase materials and foods for the foreign ships which berthed at the waters of Whampoa and Macao; the second were the merchant houses' compradors in charge of general affairs and cash for the foreign traders. The compradors evolved from the overmen who served foreign traders. Their identities were more special than that of the ordinary servants in that they could only work as compradors under the guarantee of the Hong merchants and linguists, and were given licenses by the Canton Customs. In the eyes of some foreign traders who lived in the merchant houses, the compradors were even more important and powerful

March / 2021.3

31

星期三

Wednesday

辛丑牛年 辛卯月

二月十九

than the Hong merchants, mainly because all the Chinese service personnel in the merchant houses, including foreign trade accountants, cooks, servants and coolies, were at the beck and call of the compradors.

April

一	二	三	四	五	六	日
			1 二十	2 廿一	3 廿二	4 清明
5 廿四	6 廿五	7 廿六	8 廿七	9 廿八	10 廿九	11 三十
12 三月	13 初二	14 初三	15 初四	16 初五	17 初六	18 初七
19 初八	20 谷雨	21 初十	22 十一	23 十二	24 十三	25 十四
26 十五	27 十六	28 十七	29 十八	30 十九		

十三行同文街上的店铺商人与外商。（图片采自《广州历史文化图册》）

Shop owners and foreign merchants in New China Street of the trading region. (The picture is taken from *Historical and Cultural Pictorial of Canton.*)

April / 2021.4

1

星期四

Thursday

辛丑牛年 辛卯月

二月二十

早期代理人

Intermediaries in the Early Period

买办不仅掌握着服务外商的人事大权，也管理着外商在商馆的办事机构和账目，甚至连存放所有现款和账本、文件、信函等贵重物品的商馆银库也在买办管理之下。一方面，他们照拂着外商的衣食住行，无微不至侍候着洋行大班，另一方面又执行着政府交予的管束外国人的任务。

买办虽然肩负着兼管外商的责任，但同样也很容易被外商支配。比如买办的薪资虽然少，但他的雇主却允许他从别的事项上获得大额津贴，类似在鉴定外国银元时收取手续费（每1000元抽2分）；商馆对外付款时收取“底子钱”（1000元以下每元收5文）；采办商馆必需品时可抽佣金等。可以说，买办和外国商人被金钱关系绑在了一起。到了后来，他们中间逐渐分化出一批人，代替外商出面从事投资经营活动，并帮助外商与政府周旋，成为近代买办阶级的前身。

The compradors not only took control of the personnel who served the foreign traders, but also managed the institutions and accounts of the foreign traders in the merchant houses. Even the account books, documents, letters and other valuables of the treasury were under the compradors' management. On the one hand, they took care of the foreign traders' clothing, foods, housing, and traveling, and provided the supercargoes with considerate service. On the other, they carried out the tasks assigned by the government to supervise foreigners.

Although the compradors had the responsibility of supervising the foreign traders, they might also be manipulated by foreigners. For example, the compradors' salaries were small, but their employers allowed them to get a large amount of allowance from other matters. For instance, a comprador might collect a commission fee (2 *fen* from 1000 silver dollars) from the authentication of foreign

April / 2021.4

2

星期五

Friday

辛丑牛年 辛卯月

二月廿一

silver dollars; when a certain merchant house carried out its payment, its comprador could collect 5 *wen* from every silver dollar if the total sum was below 1000 silver dollars; he could get commission when purchasing necessities for the merchant house. It could be acknowledge that the compradors and foreign traders were bound together by a commercial relationship. Later, they gradually changed into a group of people who represented the foreign traders to engage in investment and management activities, and help the foreign traders deal with the government. They became the predecessors of the modern comprador class.

(Note: *Fen* and *wen* were the currency units of that period.)

清道光广彩花鸟人物纹方形花口盘。

Plate with patterns of figures and flowers.reign of Emperor of Daoguang, Qing dynasty.

April / 2021.4

3

星期六

Saturday

辛丑牛年 辛卯月

二月廿二

清光绪广彩花鸟人物纹奶壶。

Milk pot with patterns of figures, flowers and birds, reign of Emperor of Guangxu, Qing dynasty.

清明

Tomb-Sweeping Day

April / 2021.4

4

星期日

Sunday

辛丑牛年 壬辰月

二月廿三

行商有制度 Ⅰ

The Canton System Ⅰ

十三行也指一套完整的对外贸易制度，是清政府利用特许的垄断商人集团来干预、控制和垄断对外贸易的一项商业经济制度，目的在于“以官制商，以商制夷”。十三行制度也称为“行商制度”，包括承商、保商、公行、总商、行佣等。这套制度起于康熙开海初期，并在随后数十年的对外贸易过程中逐步添加完善而成。

承商制度：指行商“承揽夷货”制度。最初行商承商，只须具备一定资产，自己向地方官府申报便可领帖营业；到了乾隆年间，在资产殷实的基础上，还必须有现任行商一至两人做保；嘉庆十八年（1813年）之后，行商准入审批权收归户部，由公行内所有行商，包括总商在内“慎选殷实公正之人”共同联名结保，呈至户部察验。此外，清政府还赋予行商一种“世袭”性质，规定不能自由辞退，即使是老弱病残无力承商，也应由其亲信子侄接办。

The “Thirteen Hongs” also referred to a complete foreign trade system, which was built by the Qing government to authorize the Hong merchant groups to own monopoly, so as to intervene, control and monopolize the foreign trade. As a commercial and economic system, its purpose was to “use the officials to control the merchants, and use the merchants to control the foreigners”. The Thirteen Hongs System, also known as the “Hong Merchant System”, which included regulations of underwriting, standing security, establishing a merchant guild, designating a general merchant, and collecting commissions. This system was implemented in the early days of the Emperor Kangxi’s reign when he lifted the ban on the maritime affairs, and was gradually improved and completed in the following decades of foreign trade.

April / 2021.4

5

星期一

Monday

辛丑牛年 壬辰月

二月廿四

The Underwriting System: The Hong merchants would "underwrite the foreign goods". At the beginning, the merchants only needed to have certain assets to declare to the local government, and apply for permission to trade. In the reign of Emperor Qianlong, in addition to owning abundant assets, the merchants who applied for approval to trade also needed the security stood by one or two Hong merchants. After the eighteenth year of Emperor Jiaqing (1813), the right to grant the merchants access was returned to the Ministry of Revenue. All the Hong merchants in the merchant guild, including the chairman, should "carefully select the rich and fair persons", and jointly stood security for the new Hong merchants. The signed guarantee would be submitted to the Ministry of Revenue for inspection. In addition, the Qing government also gave merchants a hereditary title, stipulating that they could not resign freely. Even the old, the sick and the disabled who were unable to undertake the business, should be replaced by their close relatives and nephews.

April / 2021.4

6

星期二

Tuesday

辛丑牛年 壬辰月

二月廿五

《十三行货栈图》，原件藏于大英图书馆。（图片采自《大英图书馆特藏中国清代外销画精华》）

Warehouses in the trading region. The original piece is collected in the British Library. (The picture is taken from *Chinese Export Paintings of the Qing Period in The British Library*.)

行商有制度 Ⅱ

The Canton System Ⅱ

保商制度：行商垄断了广州进出口业务，进口货物由其承销，内地出口货物由其代购，并且负责划定进出口货物的价格，以及向海关担保外国商船缴纳进出口关税，即“承保税饷”，所以行商又称为“保商”。身为保商，往往要先拿出真金白银为外国商人垫资，自己也需要缴纳税金。假如外商拖欠税款，保商要负连带责任；而相应的，保商亦对外国商船货物享有优先权利。

总商制度： 1813年，在保商制度的基础上，进一步演化出总商制度，即在保商之上又设有总商，由资历深、资产多的行商出任，其职责主要包括征收行佣、协调货价、贯彻政府相关政策法令、办理中外交涉等，并对整个行商团体负责。

The Guarantor system: The Hong merchants monopolized the import and export trade of Canton, and the imported goods were underwritten by them. Representing the foreign traders, they purchased export goods from the mainland, and were responsible for defining the price of the import and export goods and guaranteed that the foreign merchants would pay the Customs import and export duties. This was known as "responsibility for the taxation". Therefore, the Hong merchants were also called "guarantee merchants". It was often necessary for a guarantor to use gold and silver to advance money for foreign merchants. But they also needed to pay taxes. If the foreign traders defaulted on the tax payment, the guarantor would be responsible for that; the guarantors would also enjoy the priority right on the cargoes from foreign merchant ships.

The General Merchant System: In 1813, on the basis of the guarantor system, the general merchant system was further evolved. Based on the guarantor system,

April / 2021.4

7

星期三

Wednesday

辛丑牛年 壬辰月

二月廿六

there was also a position called the general merchant, whose position would be taken by a Hong merchant who was highly qualified with abundant asset. His duties mainly included collecting commissions, coordinating the price of goods, implementing the relevant government policies and decrees, dealing with foreign negotiations, and so on. He would be responsible for the whole merchant guild.

外国人眼中的广州帽店。（图片采自《昔日乡情》）

This is a Cantonese hat shop in the eyes of the foreigners. (The picture is taken from *The Nostalgia for the Hometown*.)

April / 2021.4

8

星期四

Thursday

辛丑牛年 壬辰月

二月廿七

行商有制度 Ⅲ

The Canton System Ⅲ

公行制度：保守严厉的外贸政策和巨大的外贸利益引发了不同利益集团之间的争夺。在行商之外，还出现了由不同官僚势力控制的"皇商""总督商人""将军商人"和"抚院商人"等。竞争最激烈的一次，是行商联合外商自发组织起来对抗一个"皇商"。这次抗争以"皇商"失败告终，但也让行商们意识到联合的重要性。

康熙五十九年（1720 年），公行正式成立，并订下十三条行规，在货物价格约定、责权利的划分等方面做了规定，目的在于保证相对公平、防止内部恶意竞争、垄断大宗商品贸易。然而，这些规定对于奉行贸易自由的外商来说无法容忍，对于行外商人也不利，于是两者联合起来进行反对，导致公行成立后第二年即宣告解散。

The Merchant Guild System: The conservative and strict foreign trade policies and huge interests of foreign trade had caused disputes among different interest groups. In addition to the Hong merchants, there also appeared the "emperor's merchants", "governor's merchants", "general's merchants" and "supervisor's merchants" controlled by different bureaucratic forces. The fiercest competition was a confrontation, voluntarily organized by the Hong merchants and foreign traders, against an "emperor's merchant". The struggle ended with the failure of the "emperor's merchant", and it also made the Hong merchants realize the importance of an alliance.

In the fifty-ninth year of Emperor Kangxi's reign (1720), "Co Hong", the merchant guild, was formally established and 13 rules were made in terms of the cargoes' prices, and the division of responsibilities and rights, so as to ensure

April / 2021.4

9

星期五

Friday

辛丑牛年 壬辰月

二月廿八

the relative fairness, prevent internal malicious competition and monopolize the commercial bulk trade. However, these regulations were intolerable to the foreign traders who pursued free trade, and were also unfavorable to the outside merchants (who were not included in the merchant guild). Therefore, the foreign traders and outside merchants united for an opposition, which led to the disintegration of the merchant guild in the next year after its establishment.

清中期铜胎画珐琅花卉纹提梁壶及温炉。茶壶通体白地，巧绘折枝花果；肩部环饰墨彩描金如意纹、黄彩锦地间以开光折枝花卉和蓝彩团凤纹、蓝彩回形纹各一周；提梁柄呈枝条状，把手处缠绕藤条用以隔热。提壶和温炉保存完整，形制特别，用料精细，釉色光亮，花卉纹描画细腻。这件藏品的造型可上溯约 1720 年的英国银器，随着茶叶出口贸易的繁荣，这种茶具的需求也相应增加。

Enamel handled teapot and burner with patterns of flowers, mid-Qing dynasty. The teapot is white and skillfully painted with twigs, flowers and fruits; the upper part is respectively encircled by the inky Ruyi patterns with golden decorations, and the yellow brocade decorated with twigs and flowers in the open windows, the blue phoenix patterns and the blue fretted patterns; the handle, in the shape of branches, is wrapped with vines for heat insulation. The well preserved teapot and burner are special in shape, made of fine materials, with bright glaze and delicately painted floral patterns. The shape of this piece can be traced back to the British silverware made in 1720. With the prosperity of export tea trade, the demand for this kind of tea sets had increased accordingly.

April / 2021.4

10

星期六

Saturday

辛丑牛年 壬辰月

二月廿九

民国铜胎画珐琅花卉纹蝴蝶形带盖盒。

Enameled copper covered box with patterns of flowers and butterflies, period of the Republic of China.

April / 2021.4

11

星期日

Sunday

辛丑牛年 壬辰月

二月三十

制度造辉煌

A Marvelous System

17 至 18 世纪是广州贸易的伟大时代，特别是 1757 年“一口通商”之后到 1842 年第一次鸦片战争之前的 85 年间，广州是官方认定的中国外贸中心。这样的辉煌不是仅凭十三行行商制度就能达到的，而是在官方制定规则的前提下，由引水人、通事、行商、买办等形色职业构建出一条环环相扣的贸易体制链条，对外商及贸易船只进行有效引导和管理，而针对贸易的所有监督控制权则集中于粤海关手中。粤海关监督和两广总督通过控制所有与外国商人打交道的中国人，绝对支配着贸易的发展。

The 17th and 18th century was the great era of Canton trade. In the duration of 85 years, from the one-port trade policy implemented in 1757 to 1842 when the First Opium War broke out, Canton was officially recognized as China's foreign trade center. Such a marvelous achievement was not made only by the Canton system, but also by an interconnected trading chain consisting of various professions such as pilots, linguists, Hong merchants, compradors, under the premise of official rules, which effectively implemented guidance and management on foreign traders and ships, and concentrated all the supervision rights and control over the trade in the Canton Customs. In this way, the supervisor of Canton Customs and the governor of Guangdong and Guangxi provinces controlled all the Chinese who dealt with foreign traders, and absolutely dominated the trade development.

April / 2021.4

12

星期一

Monday

辛丑牛年 壬辰月

三月初一

17至18世纪是广州贸易的黄金时代。广州是当时官方认定的中国外贸中心。（图片采自《珠江风貌》）

The 17th and 18th centuries were the golden era of Canton trade, at that time, Canton was the officially designated trading center of the Sino-Western trade. (The picture is taken from *Views of the Pearl River Delta, Macao, Canton, and Hong Kong*.)

April / 2021.4

13

星期二

Tuesday

辛丑牛年 壬辰月

三月初二

奇珍入中华

Rare Curiosities

自“一口通商”之后，广州十三行基本成为清廷西洋珍奇的唯一“供应商”。粤海关监督会亲自前往黄埔锚地丈量外商船只，一个重要目的便是挑选出稀少而有价值的珍宝货物进献给上级和皇帝。外国商人显然也深谙清朝官员心理，英国东印度公司、英法散商、亚美尼亚商人、穆斯林商人、巴斯商人等都会随船携带数量不等的西洋奢侈品。粤海关监督挑选出最好的珍品后，会要求行商购买这些东西，然后行商再以超低折扣转卖给粤海关监督。

清代“采办官物”的档案显示了皇帝对广东洋货贡品的具体指示，如乾隆极其痴迷西洋珍宝，曾在 1758 年的一道谕旨中指明“买办洋钟表、西洋金珠、奇异陈设或新样器物”。为维持承保船只的特权，行商不得不忍受来自官府的盘剥，这也是导致十三行行商集团最后没落的一大原因。

Since the one-port trade policy had been implemented, the Thirteen Hongs in Canton had basically become the only “supplier” of Western treasures in the Qing dynasty. The supervisor of Canton Customs would go to Whampoa anchorage in person to measure the foreign ships. One of the important purposes of that was to select rare and valuable treasures for the higher authorities and the emperor. Foreign traders were obviously familiar with the mentality of the officials of the Qing court. The English East India Company, the English and French private traders, the Armenian merchants, the Muslim merchants, the Parsee merchants would carry different quantities of Western luxury goods on their ships. When the supervisor selected the best curiosities, they asked the Hong merchants to buy them, and then the Hong merchants would re-sell them to the supervisor of Canton Customs at a ☞

April / 2021.4

14

星期三

Wednesday

辛丑牛年 壬辰月

三月初三

super low discount.

The archives of "the purchase of official things" in the Qing dynasty showed the emperors' specific instructions for Canton's tributes of Western goods. For example, Emperor Qianlong was extremely obsessed with Western treasures. In an edict of 1758, he pointed out the officials should "buy foreign clocks and watches, Western gold beads, rare furnishings or novel artifacts". In order to maintain the privilege of underwriting, the Hong merchants had to endure the exploitation from the government, which was also one of the reasons for the final decline of the Hong merchant groups.

粤海关官员丈量外国商船图。（图片采自 *Den långa resan*）

The Canton Customs officials were measuring the foreign merchant ships. (The picture is taken from *Den långa rensan*.)

April / 2021.4

15

星期四

Thursday

辛丑牛年 壬辰月

三月初四

粤商威名扬

Fame of the Cantonese Merchants

十三行也指由一群商人联合结成的公行组织。所谓的“十三”，并不具有数字上的规约，只是一群行商的统称。他们是“身家殷实，资财素裕”的商人，懂得各国语言，熟悉洋行商务，了解清政府的运作。这些行商的数目，最多的是乾隆二十三年（1758 年）的 26 家，最少是乾隆二十二年（1757 年）的 4 家，也有刚好 13 家的时候（嘉庆十八年和道光十七年）。

作为官方指定的对外贸易经纪人，十三行具有的垄断性贸易特权为他们带来了财富的迅速积累，所谓“粤东十三家洋行，家家金珠论斗量。楼阑粉白旗杆长，楼窗悬镜望重洋”。很快，十三行粤商便与两淮盐商、山西晋商并称为清代中国三大商人集团。道光时期曾流传谣谚称“潘卢伍叶，谭左徐杨，虎豹龙凤，江淮河汉”。其中的潘、卢、伍、叶正是十三行的首富。

The Thirteen Hongs also referred to the merchant guild formed by the Hong merchants. The so-called “thirteen” was not a numerical stipulation, but an appellation of a group of Hong merchants. They were merchants “from rich families and with abundance of wealth”. They knew several languages of different countries, and were familiar with the foreign trade, with an understanding of the operation of the Qing government. The number of these Hong merchants was 26 at most in the twenty-third year of Emperor Qianlong's reign(1758), 4 at least in the twenty-second year of Emperor Qianlong's reign(1757), and coincidentally 13 in the eighteenth year of Emperor Jiaqing's reign as well as the seventeenth year of Emperor Daoguang's reign.

As the officially appointed foreign trade intermediaries, the Hong merchants rapidly accumulated their fortunes by the monopoly privilege. There was a

April / 2021.4

16

星期五

Friday

辛丑牛年 壬辰月

三月初五

description as follows: "there are Thirteen Hongs in Canton; the gold and pearls in every merchant house are immeasurable. The balustrades of the buildings are white, while the flagpoles are long, and you can see the ocean in the mirror hanging inside a room with a window". Soon, the Cantonese Hong merchants, the salt merchants of the Huai River region and the Shanxi merchants were known as the three famous merchant groups of China in the Qing dynasty. During the period of Emperor Daoguang's reign, there was a saying claiming that "the families of Pan, Lu, Wu, Ye, and the families of Tan, Zuo, Xu, Yang has strong power like tigers and phoenixes". The families of Pan, Lu, Wu and Ye were the richest among the Hong merchants.

19世纪末菩提叶水彩人物画。

A figure, watercolour on bodhi leaf, late Qing dynasty, late 19th century.

April / 2021.4

17

星期六

Saturday

辛丑牛年 壬辰月

三月初六

19 世纪末菩提叶水彩人物画。

A figure, watercolour on bodhi leaf, late Qing dynasty, late 19th century.

April / 2021.4

18

星期日

Sunday

辛丑牛年 壬辰月

三月初七

富而好礼也

The Wealthy and Hospitable Merchants

即便拥有富可敌国的财富，在封建社会，商人的地位仍不高，所以发家之后的四大行商，几乎不约而同地将文化当作一项事业或兴趣来经营。如同文行潘氏，先后出了五位举人、四位翰林。据《番禺河南小志》记载："伍氏喜刻书，叶氏喜刻帖，潘氏独以著作传。" 伍崇曜致力于搜书、藏书、刻书，前后所辑刻书有《岭南遗书》《粤十三家集》《楚庭耆旧遗诗》《粤雅堂丛书》等。潘仕成则将海山仙馆内的数万卷藏书和千多块名匠石刻编撰刊印成《海山仙馆丛书》，共 56 种 485 卷。伍崇曜的粤雅堂和潘仕成的海山仙馆藏书甚巨，与康有为的万木草堂、孔广陶的岳雪楼并称"粤省四家"（即四大藏书楼）。行商们对于陶瓷、绘画、家具、服饰、摆设、生活模式的爱好，一度成为当地文人雅士和普罗大众所追逐的潮流风向标。

Though with a fortune amounting to that of a country, the merchants were still in a low status in the feudal society, so after the four merchant families had earned their fortune, they, in sync with each other, took culture as their enterprise or interest. For example, five Jurens (successful candidates in the imperial examination at the provincial level) and four Hanlins (members of the imperial academy) came successively from the Pan family. *The Records of the Chronicles of the Southern River in Panyu* mentions that "the Wu family likes to publish books, the Ye family likes Chinese calligraphy, and the Pan family is famous for their writings". Wu Chongyao was devoted to searching, collecting and publishing books. The books successively published by the Wu family include *The Bequeathed Book of Lingnan*, *Works of Thirteen Cantonese Witters*, *The Old Poems of the Chu Court*, and *The Serial Books of Yueyatang, Hall of Elegance in Canton.* Pan Shicheng compiled

April / 2021.4

19

星期一

Monday

辛丑牛年 壬辰月

三月初八

tens of thousands of volumes of books and the contents of thousands of stone carvings created by famous craftsmen and printed them into *the Serial Books of Haishanxianguan*, the Garden of Immortals on the Sea and Mountain, which includes 56 types and 485 volumes. There was a huge collection of books in Wu Chongyao's Yueyatang (Hall of Elegance) in Canton and Pan Shicheng's Haishanxianguan(Garden of Immortals on the sea and Mountain). They were also known as the "four libraries in Guangdong province" together with Kang Youwei's Wanmucaotang (Hall of Trees and Grass) and Kong Guangtao's Yuexuelou (Tower of the Mountain Snow). For a time, the Hong merchants' interest in porcelain, paintings, furniture, clothing, furnishings and life style became the trend and fashion pursued by the local literati and the general public.

谷雨 Grain Rain

April / 2021.4

20

星期二

Tuesday

辛丑牛年 壬辰月

三月初九

伍秉鉴肖像画。伍秉鉴（又称浩官）为十三行怡和行行主，也是伍崇曜的父亲。他德高望重，广交美国商人，并投资他们的企业。（图片采自《清代洋画与广州口岸》）

A portrait of Wu Bingjian. Wu Bingjian (also known as Howqua), father of the Wu Chongyao, was the owner of Yihe Hong. He had noble character and enjoyed high prestige, and well-connected with the American merchants, in whose enterprises he also made investments. (The picture is taken from *Western Paintings and the Canton Port during the Qing Period*.)

潘振承：商总第一人

Pan Zhencheng: the Most Famous Hong Merchant

潘振承，又名潘启，出生于福建泉州一个贫苦农民家庭，年少时便到海上当船工，三下吕宋经商，学会了西班牙语、葡萄牙语和英语。之后，潘振承来到广州，在一个闽籍陈姓行商的洋行里做事，经办洋行一切商务，深得信任。几年后，陈老板停业回乡，潘振承便创立了同文洋行。由于具有语言优势，加上做生意诚信，潘振承很快被外国商人誉为“最可信赖的商人”，几乎垄断了十三行与英国的生丝对外贸易，是英国东印度公司最大的客户，也是瑞典东印度公司最主要的贸易伙伴。乾隆二十五年（1760 年），潘振承被选为十三行商总，潘家成为连续出任商总时间最长的家族，长达 39 年。

Pan Zhencheng, also known as Pan Qi, was born in a poor peasant family in Quanzhou, Fujian province. When he was young, he worked as a boatman at sea. He had been to Lusong (now an island in the Phillipines) to trade and learned Spanish, Portuguese and English. After that, Pan Zhencheng came to Canton and worked in a merchant house whose owner was from Fujian and bore the family name Chen. He was deeply trusted to handle all businesses in the merchant house. A few years later, when his boss Chen gave up trading and went back to his hometown, Pan Zhencheng founded his merchant house, Tongwen Hong. Because he had the advantage in language and highly credited in business, Pan Zhencheng was soon recognized as “the most reliable merchant” by foreign traders. He almost monopolized the foreign trade of raw silk between the Hongs and the UK. He was the most important client of the English East India Company and also the main trading partner of the Swedish East India Company. In the twenty-fifth year of Emperor Qianlong's reign (1760), Pan Zhencheng was elected as the general merchant of the merchant guild. The Pan

April / 2021.4

21

星期三

Wednesday

辛丑牛年 壬辰月

三月初十

family was the family who had taken the position of the chairman of the merchant guild for as long as 39 years, which was the longest tenure.

April / 2021.4

22

星期四

Thursday

辛丑牛年 壬辰月

三月十一

潘振承玻璃肖像画。瑞典“哥德堡号”来广州与潘振承进行贸易时，潘振承赠送了一幅自己的玻璃画像。至今这幅画像仍保留在哥德堡市的博物馆里，是欧洲所有博物馆中珍藏的唯一一幅中国人画像。（图片采自《广东十三行与早期中西关系》）

The portrait of Pan Zhencheng. When the Swedish ship “Gotheborg” came to Canton to trade, Pan Zhencheng presented his own portrait, a reverse glass painting, to the merchants as a gift. Now the portrait is collected by a museum in Gotheborg. (The picture is taken from *The Thirteen Hongs and the Sino-Western Relationship in the Early Period.*)

伍秉鉴：投资美利坚

Wu Bingjian: Invest in the United States

2001 年，美国《华尔街日报》统计了 1000 年来世界上最富有的 50 人，有 6 名中国人入选，其中就包括十三行怡和行的主人伍秉鉴。

伍家祖上是福建茶农，到了伍秉鉴父亲这一代创立了怡和行，主要从事茶叶、丝织品和瓷器的对外贸易。伍秉鉴具有超前的经营理念，每年与英美商人贸易额高达数百万银元。怡和行的茶叶曾被英国公司鉴定为最好的茶叶，市场价格颇高。伍秉鉴精于投资，不仅在国内投资了地产、房产、茶园、店铺等产业，还隔着重洋委托自己在美国的代理人投资铁路、证券和保险等业务，泛美大铁路、密歇根中央铁路、密苏里河铁路等都曾是伍秉鉴的投资对象。怡和行成了一个名副其实的跨国财团。从嘉庆十四年 (1809 年) 起，英国驻华历任商务监督，都必须与伍家打交道。

In 2001, the *Wall Street Journal* of the United States counted the world's 50 richest people in the past 1000 years. Six Chinese were counted in, including, Wu Bingjian, the owner of Yihe Hong.

The ancestors of the Wu family were Fujian tea farmers. In the generation of Wu Bingjian's father, Yihe Hong was founded, mainly engaged in foreign trade of tea, silk and porcelain. Wu Bingjian had an advanced idea towards trading, and the annual trade volume with British and American traders was up to millions of silver dollars. Yihe Hong's tea was identified as the best by English East India Company, and its market price was pretty high. Wu Bingjian, good at investment, not only invested in land, real estate, tea gardens, shops and other industries in China, but also entrusted his agent in the United States to invest in railway, securities, insurance and other businesses across the ocean. The Pan-American Railway, Michigan Central Railway

April / 2021.4

23

星期五

Friday

辛丑牛年 壬辰月

三月十二

and Missouri River Railway were all once Wu Bingjian's investment targets. Yihe Hong had truly become a transnational consortium. Since the fourteenth year of Emperor Jiaqing's reign（1809）, all the British business supervisors in China must deal with the Wu family.

清末人物船舶图菩提水彩画（之一）。这组绘画共6 幅，以菩提叶为材料，水彩绘制船舶、市井人物、乐手演奏图。此幅菩提水彩画中的人物描绘生动，神态自若；船舶描绘细腻，方寸之间的画面色彩明暗处理得当，富有立体感。菩提叶是除通草纸外，一种成本相对低廉的材料，在清代广州外销画中占据着一席之地。

Ships and figures, watercolour on bodhi leaf, late Qing dynasty. There are 6 paintings in this set, with bodhi leaves as materials, and themes like ships, people in the markets and musicians playing the instruments are painted in watercolours. The figures in this set of paintings on bodhi leaves are vividly depicted with a self-contained manner; the ships are depicted delicately, and the colours in detail and the contrast between shadow and brightness are properly handled with a three-dimensional effect. As relatively low-cost materials in addition to the pith paper, bodhi leaves also played a part in the Cantonese export paintings in the Qing dynasty.

April / 2021.4

24

星期六

Saturday

辛丑牛年 壬辰月

三月十三

April / 2021.4

25

星期日

Sunday

辛丑牛年 壬辰月

三月十四

19世纪玻璃画仕女图。画中的仕女手拮灵芝，极具东方色彩的吉祥寓意。玻璃画是外销画的重要画种之一。仕女的绘画风格与其服饰、发饰等时代特征明显，对于外销玻璃画的研究具有比较重要的参考价值。

A maiden, reverse glass painting, 19th century. The maiden in the painting is holding a ganoderma lucidum, which has auspicious implications and an oriental aura. The reverse glass paintings were one of the most important export paintings. The painting style, dresses and hair ornaments of the ladies were obviously in tune with the era, which is of great referential value for the study of export reverse glass paintings.

卢观恒：大器晚成者

Lu Guanheng: an Extraordinary Man Succeed Late in Life

在十三行有名的行商中，虽不乏出身贫寒的人（如潘振承），但年过四十依然一无所有，尔后峰回路转，短短几年富甲天下的唯有卢观恒一人。

卢观恒四十多岁的时候，离开家乡新会来到广州谋生，帮人看守歇业的店铺。当时有外商在贸易季结束时将没有卖出的货寄放在卢观恒看守的空铺中，委托他代为销售。没想到，完全没有经商经验的卢观恒居然将货物全部售罄，令外商刮目相看，被升为洋行买办，开始接触、学习外贸业务。1792 年，卢观恒正式充任行商，创立广利行，主要从事茶叶和棉花贸易。在卢观恒的经营下，广利行迅速发展，1796 年居行商第三位，第二年跃居第二位，更于 1800 年与同文行潘振承平起平坐，同为行商首领。

Though many of the Hong merchants came from poor backgrounds (such as Pan Zhencheng), Lu Guanheng was the only one who had not achieved success before his forties. His way to achieve insurmountable fortune was full of twists and turns.

When Lu Guanheng was in his forties, he left his hometown Xinhui and came to Canton to make a living, serving as a guard to watch the closed shops. At that time, some foreign traders consigned the unsold goods to the empty shop guarded by Lu Guanheng at the end of the trading season and entrusted him to sell them. Unexpectedly, Lu Guanheng, who had no business experience at all, sold out all the goods, which impressed the foreign traders so much. He was then promoted to be a comprador and began to learn and deal with the foreign trade affairs. In 1792, Lu Guanheng officially became a trader and founded Guangli Hong, mainly engaging in tea and cotton trade. Under Lu Guanheng's management, Guangli Hong developed

April / 2021.4

26

星期一

Monday

辛丑牛年　壬辰月

三月十五

rapidly. In 1796, Lu Guanheng ranked the third place among all the Hong merchants. In the next year, he leaped to the second place and in 1800, he became the general merchant of the merchant guild on an equal footing with Pan Zhencheng, the owner of Tongwen Hong.

April / 2021.4

27

星期二

Tuesday

辛丑牛年 壬辰月

三月十六

图为卢观恒次子卢文锦肖像画。卢观恒逝世后由卢文锦承充行商，掌管行务，商名卢隶荣，外国人称其为茂官第二。（图片采自《异趣同辉》）

Lu Wenjin was the second son of Lu Guanheng. After Lu Guanheng passed away, he became a Hong merchant, in charge of the merchant house's trade. His trade name was Lu lirong and foreigners called him Maoqua II. (The picture is taken from *Chinese Export Fine Art in the Qing Dynasty*.)

潘长耀：告状到白宫

Pan Changyao: File an Appeal to the White House

潘长耀是十三行丽泉行的主人，主要从事信贷贸易，即先赊货，再结款。这种方式为他赢得很多美国贸易伙伴，但同时也带来了被拖欠货款的风险。在欠债问题上，清政府采用双重标准：若是行商欠外商债务须加倍偿还；若是外商欠行商债务则不闻不问。1804 年，因美国商人欠债不还，粤海关监督和两广总督互相推诿不作为，潘长耀前往美国宾夕法尼亚东部联邦法院起诉其里斯、格里夫斯和米福临公司，要求他们偿还 2.5 万西班牙银元债务。这是中国第一起跨国民事诉讼案件。官司在反复的控告与反控告之间一直持续到 1813 年，丽泉行已经濒临倒闭。第二年 2 月，潘长耀直接写信给美国总统麦迪逊提出申诉，依然未果。这封信如今保存在美国国家档案馆内。此后，潘长耀又几次向美国法院提出诉讼，然而，直到 1823 年丽泉行破产，仍未收回全部欠款。

Pan Changyao, the owner of Liquan Hong, was mainly engaged in credit trade. The merchandise was obtained on credit first, and then the bill would be paid afterwards. In this way, he had won over many American trading partners, but also had to take the risk of being defaulted. Toward the issue of debt, the Qing government adopted a double standard: if the Hong merchants owed debts to the foreign traders, they had to double the payment; if the foreign traders owed debts to the Hong merchants, the government would simply ignore that. In 1804, the American traders failed to repay their debts, but the supervisor of Canton Customs and the governor of Guangdong and Guangxi provinces deflected responsibility to each other and ignored the problem. Pan Changyao then went to the Federal Court in the East of Pennsylvania, United Sates, to sue three American companies, asking them to repay the debt of 25,000 Spanish silver dollars. This was the first

April / 2021.4

28

星期三

Wednesday

辛丑牛年 壬辰月

三月十七

transnational lawsuit in China. Between repeated accusations and counter-accusations, the lawsuit lasted until 1813, when Liquan Hong was on the verge of bankruptcy. In February next year, Pan Changyao wrote directly to the U.S. President James Madison to file an appeal, but still failed. The letter was now in the National Archives of the United Sates. After that, Pan Changyao also filed several lawsuits to the Courts of the United States. However, the debts had not been fully repaid by 1823 when Liquan Hong went bankrupt.

十三行丽泉行主人潘长耀修建的豪宅。（图片采自《昔日乡情》）

The grand villa built by Pan Changyao, the owner of Liquan Hong. (The picture is taken from *The Nostalgia for the Hometown.)*

April / 2021.4

29

星期四

Thursday

辛丑牛年 壬辰月

三月十八

潘仕成：富贵云烟散

Pan Shicheng: All the Fortune Had Gone

岭南第一名园海山仙馆的主人潘仕成，其先祖以盐商起家，给他留下了丰厚的家底。他先盐务入洋务，成为广州一代巨商；再由商入仕，历任两广盐务使，浙江盐运使、布政使，职衔达到从二品，是晚清享誉朝野的官商巨富。

潘仕成是出色的商人，更是手腕出众的外交人才，是清政府对外事务的顾问，协助广东各级官员参与多次外交活动。他既喜好享乐风雅，玩出了一种以其姓命名的紫砂壶形制，也急公好义，办学、修路、赈灾、种牛痘，无不慷慨解囊。他是大众慈善家、军事开发的财阀、中央财政的无私资助者，然而这种种都需要庞大的资金作为后盾。由于经营不善，加上各方面开支负担过重，到了潘仕成晚年，潘家因财务问题横遭变故，家产变卖殆尽，连海山仙馆也被没官拍卖。潘仕成在穷困潦倒中死去。

Pan Shicheng was the owner of Haishanxianguan (the Garden of the Immortals on the Sea and Mountain), the most famous garden in Lingnan region. His ancestors started their business as salt merchants, leaving a large amount of fortune to him. He quit the salt trade, engaged in foreign trade, and became a famous Hong merchant in Canton. He even went into politics, successively serving as the administrator of Guangdong and Guangxi provinces on salt trade, administrator of Zhejiang province on salt transportation, and Buzhengshi (chief director of provincial political affairs) with the second grade ranking. He was a famous and wealthy official merchant in the late Qing dynasty.

Pan Shicheng was an outstanding merchant, diplomat, and also an adviser to the Qing government on foreign affairs and assisted officials at all levels in Guangdong province to participate in many diplomatic activities. Not only did he enjoy pleasure

April / 2021.4

30

星期五

Friday

辛丑牛年 壬辰月

三月十九

and elegance, from which appeared a kind of dark-red enameled pottery named after him, but was also keen on charity such as building schools and roads, making donations to the stricken areas, and patronizing the vaccination against smallpox, to which he contributed generously. He was a public philanthropist, a plutocrat of military development, and a selfless supporter of the central government. However, these activities needed to be backed by huge funds. Because of poor management and overburdened expenditure in all aspects, in Pan Shicheng's later years, Pan's family suffered from financial problems, and their property was sold out. Even his garden was confiscated and auctioned. Pan Shicheng died in poverty.

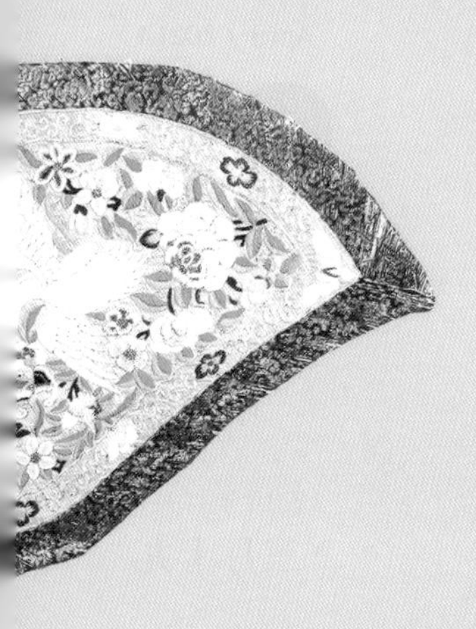

May

一	二	三	四	五	六	日
					1 劳动节	2 廿一
3 廿二	4 五四青年节	5 立夏	6 廿五	7 廿六	8 廿七	9 廿八
10 廿九	11 三十	12 四月	13 初二	14 初三	15 初四	16 初五
17 初六	18 初七	19 初八	20 初九	21 小满	22 十一	23 十二
24 十三	25 十四	26 十五	27 十六	28 十七	29 十八	30 十九
31 二十						

清中期青花五层楼图八角碟。

Blue and white dish with illustration of the five-storied pagoda, mid-Qing dynasty.

May / 2021.5

1

星期六

Saturday

辛丑牛年 壬辰月

三月二十

劳动节

International Worker's Day

May / 2021.5

2

星期日

Sunday

辛丑牛年 壬辰月

三月廿一

清乾隆广彩珍珠地庭院人物纹双龙耳狮钮盖对瓶。

Vases with patterns of gardens and figures on pearl ground, reign of Emperor of Qianlong, Qing dynasty.

叶上林：急流知勇退

Ye Shanglin: Retired at the Peek of His Career

义成行主人叶上林，早年因家贫进入潘家的同文行做账房先生，深得器重。乾隆五十七年（1792 年），叶上林离开同文行，与行商石中和合伙做外贸生意。没过几年，因资金链断裂破产，叶上林也背上了不少债务。就在这时，一位外商朋友替他还掉几笔最要紧的债务，叶上林因此渡过了难关，并创立了义成行。

义成行创立后没多久，十三行迎来了一波破产潮，叶上林趁机接手了一些资质不错的破产小行商及他们的贸易份额，迅速坐大。19 世纪初，叶上林已经拥有多个商号，从事各种货物的买卖，俨然一个大型进出口贸易集团，而他却在此时慢慢收缩市场份额，直至彻底退出商界，以叶廷勋（花溪老人）之名，开始了写诗读书的文人富家翁生活。也正是叶上林的退休，让叶家避开了十三行的衰退狂潮，成为覆巢下的完卵。

Ye Shanglin was the owner of Yicheng Hong. To get rid of his family's poverty, he worked for Tongwen Hong of the Pan family as an accountant and was highly appreciated by his employer. In the fifty-seventh year of Emperor Qianlong's reign (1792), Ye Shanglin left Tongwen Hong and cooperated with Shi Zhonghe, a Hong merchant, to conduct foreign trade. In a few years, Ye Shanglin owed a lot of debts because of the broken capital chain. At that time, a foreign friend paid off some of his most urgent debts, and Ye Shanglin got through the difficulty and founded Yicheng Hong.

Not long after the establishment of Yicheng Hong, the merchant houses were overwhelmed by a wave of bankruptcy. Ye Shanglin took the opportunity to take over some small Hongs, which were qualified but went bankrupts, together with their share of trade, and quickly gained a large share in the market. By the beginning

May / 2021.5

3

星期一

Monday

辛丑牛年 壬辰月

三月廿二

of the 19th century, Ye Shanglin had already owned many brands and engaged in trades of various goods, forming a large import and export trade group. However, it was at this time that he gradually contracted his market share until he completely withdrew from the commercial world. Using the name Ye Tingxun, he began to enjoy a rich life as a man of letters, writing poems and reading books. It was Ye Shanglin's retirement that made the Ye family avoid the ferocious recession of Canton trade, and preserved themselves well in a disaster.

May / 2021.5

4

星期二

Tuesday

辛丑牛年 壬辰月

三月廿三

五四青年节

China's Youth Day

叶上林肖像画。福建省诏安县人叶上林，世居广州西关十六甫及泮塘，是十三行义成行创立者。

A portrait of Ye Shanglin. Ye Shanglin, whose ancestors were from Zhaoan town, Fujian province, lived in Shiliufu and Pantang in Xiguan area, Canton, and he was the founder of Yicheng Hong.

颜时瑛：雨打风吹去

Yan Shiying：Wrecked By Misfortune

颜时瑛的父亲颜亮洲，是泰和行的创始人，在雍正年间已是十三行行商之一。颜时瑛接管泰和行后，由于经营得法，贸易和资产规模迅速扩大。乾隆年间，泰和行已上升到十三行第三位。据史料记载，1768 年至 1779 年的行商“财富排行榜”中，颜氏位居第二。

颜时瑛还以构建磊园而闻名广州。相传磊园位于离西关绣衣坊不远的十八甫，颜时瑛耗费巨资，将它改建为拥有十八景的巨大园林，其规模之大，景致之美，在当时西关富商的园林中数一数二。颜时瑛生活奢靡，又极好客，磊园日日笙歌，达官显贵川流不息。

泰和行兴盛达四十余年，于乾隆四十五年轰然崩塌。由于挥霍无度，颜时瑛拖欠外商巨额货款，泰和行以诓骗罪名被封，颜时瑛也被革去职衔，充军伊犁，最终客死他乡。

Yan Shiying's father, Yan Liangzhou, the founder of Taihe Hong, was one of the Hong merchants in the period of Emperor Yongzheng's reign. After Yan Shiying took over Taihe Hong, the scale of trade and assets expanded rapidly due to his proper management. During the reign of Emperor Qianlong, Taihe Hong ranked the third among all the Hongs. According to historical records, the Yan family ranked the second in the ranking list of the Hong merchants' wealth from 1768 to 1779.

Yan Shiying was also famous for building Lei Garden in Canton. It was said that Lei Garden was located in Shibafu, not far from Xiuyi Lane in Xiguan. Yan Shiying spent a lot of money to transform it into a huge garden with 18 landscape scenes. The scale was so large and the landscapes were so beautiful that the garden was among the tops of the most famous gardens in Xiguan at that time. Yan Shiying

立夏

Beginning of Summer

May / 2021.5

5

星期三

Wednesday

辛丑牛年 癸巳月

三月廿四

lived a luxurious life and was extremely hospitable, and thus there would always be musical performance in Lei Garden, where numerous high officials and rich people gathered.

Taihe Hong flourished for more than 40 years and collapsed in the forty-fifth year of Emperor Qianlong's reign. Because of his extravagance, Yan Shiying owed a large debt to foreign traders. Because of fraud, Taihe Hong was closed down, and Yan Shiying was removed from his post and sent to Yili, Xinjiang, where he died as a stranger in the end.

十三行富商在西关兴建的庭院。

The villa built by the Hong merchants in Xiguan area.

May / 2021.5

6

星期四

Thursday

辛丑牛年 癸巳月

三月廿五

梁经国：逆势而上者

Liang Jingguo: Going up Against the Currents

梁经国生于1761年，广州黄埔村人，早年丧父，家境贫寒。18岁时，他入冯氏洋行做伙计。由于诚实守信，好学能干，在冯氏的资助下，他于1808年取得行商资格，创办天保行。

天宝行创立之时，正值中英关系恶化，行商纷纷破产，艰难困苦中，梁经国坚持“实在诚信”的经营之道，逐渐在与东印度公司的毛织品贸易和茶叶贸易中崭露头角，并利用良好信誉，承保东印度公司的商船，这使得天宝行的贸易地位不断上升。

1811年，梁经国与其他两个行商捐资创建文澜书院。67岁时，他退出天宝行行务，随着十三行的衰落，天宝行的后人陆续走上从政或从学之路。

Liang Jingguo was born in 1761 in Whampoa village, Canton. His father died in his early years and his family was poor. At the age of 18, he joined the Feng family's Hong as a clerk. Due to his honesty, trustworthiness, the ability to learn, as well as the Feng family's support, he obtained the qualification to be a Hong merchant in 1808 and founded Tianbao Hong.

When Tianbao Hong was founded, the relationship between China and Britain was deteriorating, and the Hong merchants went bankrupt one after another. Facing difficulties and hardships, Liang Jingguo adhered to the principle of "honesty and credibility" in his management, and gradually became a prominent figure in the trade of woolens and tea with English East India Company, and taking advantage of his good reputation, he underwrote the merchant ships of English East India Company, continuously improving Tianbao Hong's status in the trade.

In 1811, Liang Jingguo and two other Hong merchants donated money to

May / 2021.5

7

星期五

Friday

辛丑牛年 癸巳月

三月廿六

establish Wenlan Academy. At the age of 67, he quit Tianbao Hong. With the decline of the Canton trade, the descendants of Tianbao Hong began to pursue political or academic careers.

May / 2021.5

8

星期六

Saturday

辛丑牛年 癸巳月

三月廿七

清乾隆广彩锦地开光人物纹镶铜座小花盆。

Small flowerpot with copper stand and patterns of figures, reign of Emperor of Qianlong, Qing dynasty.

清乾隆广彩描金人物故事图盖盅。

Covered tureen with gold-painted patterns of figures and stories, reign of Emperor of Qianlong, Qing dynasty.

May / 2021.5

9

星期日

Sunday

辛丑牛年 癸巳月

三月廿八

火烧十三行

Conflagration in the Trading Region

在十三行时期里，曾发生过三次很大的火灾，其中最后一把火彻底烧掉了十三行。1822 年，十三行附近一家饼店失火，波及十三行，大火燃烧了两天，11 家洋行被烧掉 6 家，牵连周边民居、店铺千余间。第一次鸦片战争之后，100 多名英国士兵抢劫一家水果店并划伤店主的恶行，让愤怒的民众半夜火烧英国商馆，大火直到第二天才被扑灭，这是第二次大火。1856 年，第二次鸦片战争爆发后，驻扎在十三行地区的英军，拆毁了十三行地区周围大片民居，留下一片空地以防止中国军民的偷袭。愤怒的广州人民点燃了大火，这在历史上被称作“西关大火”。

In the Canton trade period, there had been three conflagrations, the last of which completely destroyed the trading region. In 1822, a fire broke out in a cake shop near the trading region, and spread to the merchant houses. The fire lasted two days. Six of the eleven merchant houses were burned down, while more than one thousand shops in the neighbourhood were involved. The second huge fire happened after the First Opium War, when more than 100 British soldiers robbed a fruit shop and maliciously scratched the shop owner, causing angry people to burn the English merchant house in the middle of the night, and the fire was put out the next day. In 1856, after the outbreak of the Second Opium War, the British troops stationed in the trading region demolished a large area of houses around the neighbourhood in order to create a vacant area to prevent the sneak attack by the Chinese army and people. Hence, the angry Cantonese people set fire to the region, and this was known as the “conflagration in Xiguan area”.

May / 2021.5

10

星期一

Monday

辛丑牛年 癸巳月

三月廿九

1822 年，十三行遭遇第一次大火。火势凶猛，蔓延数日，大量财物被毁于一旦。大火过后，出现了一个奇特的现象，“洋银熔入水沟，长至一二里”，真实地反映出当时行商的富裕程度。（图片采自《珠江风貌》）

In 1822, the first conflagration happened in the trading region. The situation was severe and the fire lasted a few days, destroying a large amount of fortune and goods. After the huge fire, there appeared a strange phenomenon: “the foreign silver was melted in the water and formed a stream of a kilometer or two”. It truly reflected how rich the merchants were at that time. (The picture is taken from *Views of the Pearl River Delta, Macao, Canton, and Hong Kong.*)

May / 2021.5

11

星期二

Tuesday

辛丑牛年 癸巳月

三月三十

转战上海滩

Moving to Shanghai

五口通商之后，广州的独占优势不再，大批外商纷纷转向更接近生丝和茶叶产地的宁波、上海。嗅觉灵敏的十三行行商也随着外商转移到了上海，又带动了一批为商行做事的广东人来到上海，形成了上海第一波移民潮。

广东行商到了上海后，买下大片地产，计划打造成上海版“十三夷馆”，租给外国商人。但由于清政府官员收受了英国商人的贿赂，广东行商虽名义上是地产主人，却被剥夺了对房地产的处分权，在收不到租金的情况下，被迫永久租给英国人。依托这块租界，英国商人在上海站稳脚跟，并一步步扩张，最后形成了举世闻名的外滩。

After the five ports had been open for trade, Canton no longer enjoyed the unique privilege, and a large number of foreign traders turned to Ningbo city and Shanghai city, which were closer to the production center of raw silk and tea. With foreign traders moving to Shanghai, a group of Cantonese Hong merchants together with those who worked for them also went to Shanghai, forming the first migration wave in Shanghai.

After arriving at Shanghai, the Cantonese Hong merchants bought a large amount of real estate and planned to build a Shanghai version of the “merchant houses” which would be rented by foreign merchants. However, because the Qing government officials were bribed by the British merchants, the Cantonese Hong merchants only nominally owned the real estate, but were actually deprived of the right to take control of the properties, and forced to permanently rent the merchant houses to the British traders without receiving any rent. Relying on this concession, the British traders had established their foothold in Shanghai, gradually expanded, and eventually, the Bund of Shanghai became famous all over the world.

May / 2021.5

12

星期三

Wednesday

辛丑牛年 癸巳月

四月初一

炮火下的十三行商馆区。1841年3月，英军再次北上，连陷十余座清军炮台，重占十三行街区，在夷馆再次升起英国国旗，兵临广州城门。（图片采自《广州百年沧桑》）

The trading region was bombarded. In March, 1841, the British troops marched north again, destroying more than 10 fortresses and re-occupied the trading region. The British flag was risen up again in front of the merchant house. (The picture is taken from *Vicissitude of One Hundred Years in Canton.*)

May / 2021.5

13

星期四

Thursday

辛丑牛年 癸巳月

四月初二

告别十三行

Farewell to the Thirteen Hongs

第二次鸦片战争爆发后，十三行彻底消亡——不仅指行商制度的崩塌，也指商馆区的消失。表面上是一把大火烧掉了十三行，实际上行商制度和各级官吏的压榨弊病久矣，拖垮整个行商阶层只在早晚。“一口通商”虽然让行商尽享先占优势，但与保甲制度无异的保商制度，无穷尽的纳贡、摊捐和勒索，各种债权与债务罗网，早就令行商负枷前行，举步维艰，加上英国为了扭转贸易逆差悍然走私鸦片以及因此爆发的两次鸦片战争、随之“五口通商”取代“一口通商”、中国在世界政局中的地位变化等种种因素，终于使存在了160余年的十三行宣告终结。

After the outbreak of the Second Opium War, the Thirteen Hongs completely came to an end——this not only refers to the collapse of the Canton system, but also the disappearance of the trading region. It might seem that the merchant houses were burned down in a huge fire, but in fact, the disadvantages of the Canton system and the oppression from all levels of officials had lasted for a long time and the collapse of the Hong merchant class was bound to happen sooner or later. Although the one-port trade policy allowed the Hong merchants to take advantage of all aspects, the guarantor system which was almost the same as the Baojia system (a system of taxation originated in the Song dynasty to ensure that the tax would be paid by the guarantors), the endless tributes paid to the court, donation, extortion, various obligatory rights and debts had already put the Hong merchants in shackles and trapped them in difficulties. Worse still, the British traders blatantly smuggled opium to reverse the trade deficit and as a result, led to two Opium Wars, after which, the one-port trade was replaced by five-port trade, and China's

May / 2021.5

14

星期五

Friday

辛丑牛年 癸巳月

四月初三

status changed in the World's political situation. All these factors finally brought the Thirteen Hongs of more than 160 years to an end.

清末仙鹤蝙蝠披领。

Cape with patterns of cranes, late Qing dynasty.

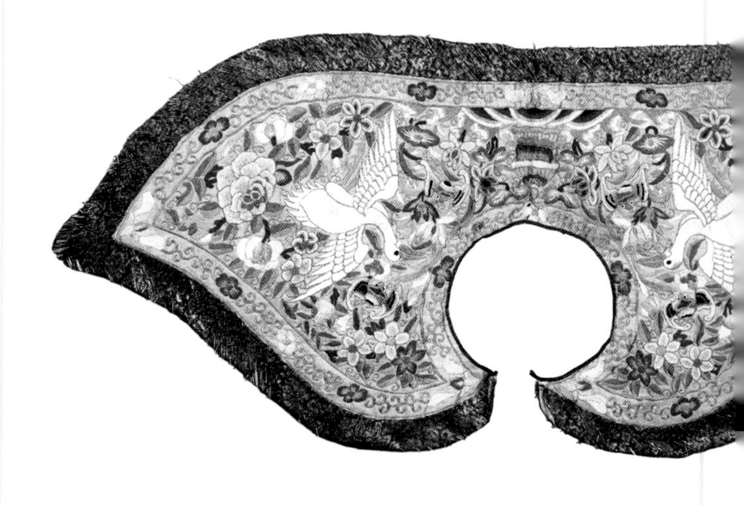

May / 2021.5

15

星期六

Saturday

辛丑牛年 癸巳月

四月初四

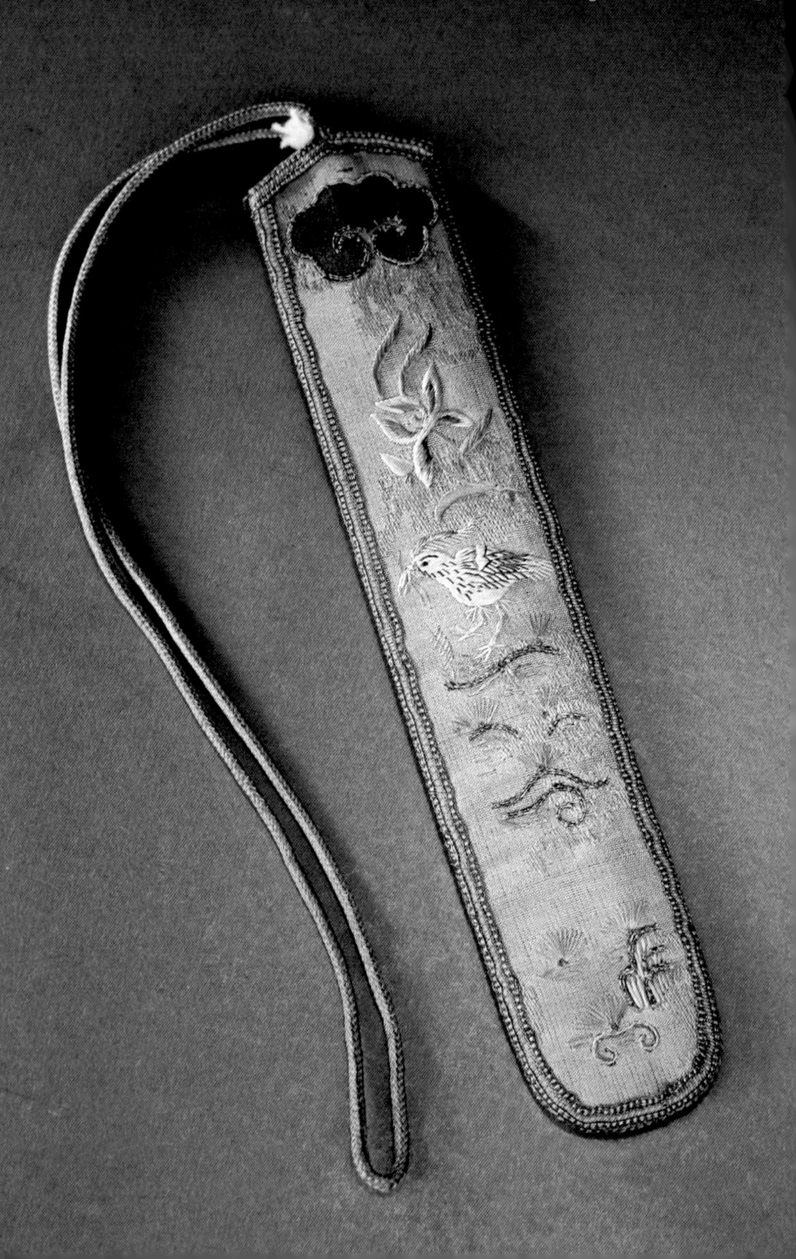

May / 2021.5

16

星期日

Sunday

辛丑牛年 癸巳月

四月初五

清末广绣亭阁花鸟扇袋。

Canton embroidery fan pouch with patterns of flowers and pavilion, late Qing dynasty.

文化交汇地

The Intersection Point of Cultures

清代“防夷五事”的出台，使十三行成为中西文化的集中交流地。清代来中国的外国人，除了商人之外，还有仰慕东方文明的科学家、艺术家、医生，以及动机复杂的传教士。他们只有通过广州行商的引见，才能得到官府的报送，获取皇帝批准的进京指令。在此之前，十三行就是他们休养身体、学习中华文化的场所。

康熙皇帝曾在1710年嘱咐两广总督赵弘灿：“西洋新来之人且留广州学汉话，若不会汉话，即到京里亦难用他。” 刚踏上天朝国土的洋人，会被安排到商馆区内的八处天主教堂学习汉语。清代宫廷十大画家之一、意大利人郎世宁和德国天文学家戴进贤等皆是在十三行接受系统中华文化培训后进入宫廷供职。

The introduction of “Five Precautions on Foreigners” in the Qing dynasty indicated that the Thirteen Hongs had become a center of cultural exchange between China and western countries. Besides foreign merchants, there were also scientists, artists, physicians who admired oriental civilization and missionaries with complicated purposes. Only through the introduction of the Cantonese Hong merchants could they get access to the government as well as permission from the emperors to go to Beijing. Before that, the trading region was the place for them to rest and study Chinese culture.

In 1710, the Emperor Kangxi told Zhao Hongcan, the governor of Guangdong and Guangxi provinces, that “the newcomers from the Western countries should stay in Canton to learn Chinese. If they could not speak Chinese, it would be difficult

May / 2021.5

17

星期一

Monday

辛丑牛年 癸巳月

四月初六

to hire them to work for the court". Foreigners who had just set foot in the kingdom would be arranged to study Chinese in eight chapels in the trading region. One of the top ten court painters in the Qing court, Giuseppe Castiglione, an Italian, and the German astronomer, Ignatius Kgler, all served the court after receiving systematic Chinese cultural training in the trading region.

一幅绘于 1860 年的画作：从珠江眺望沙面。

Overlooking Shamian Island on the Pearl River, painting, 1860.

May / 2021.5

18

星期二

Tuesday

辛丑牛年 癸巳月

四月初七

“开眼”先行人

The Open-Minded Predecessors

广东地区历史悠久的对外贸易传统，以及在行商制度下与来华外国人的深度接触，令十三行行商的眼界远远高于同时代的人，成为中国第一批“开眼看世界”的人。他们意识到中西科技方面的差距，积极出钱出力试图改善局面。如鸦片战争期间，十三行商人率先捐资引进西方先进技术，提升广东水师的装备；行商潘世荣致力于仿制当时世界上最先进的火轮船；潘仕成高薪聘请美国海军军官仿制出中国最早的近代化水雷“攻船水雷”；会隆行商郑崇谦积极推行施种牛痘，是最早传播牛痘法的中国人；伍氏家族则在将西医引入中国方面贡献良多，中国最早接种牛痘的医院“种牛痘局”和彼得·伯驾的“眼科医局”都得到了伍氏等家族的大力支持。可以说，十三行商人是“师夷长技以制夷”的最早实践者，比官方的洋务运动还早了 20 年。

Guangdong had a long history of foreign trading tradition. Under the Canton system, the Hong merchants had close contacts with foreigners who came to China, so they and insights far better than other people of the same era, and became the first group of people in China to “open their eyes to see the world”. They were aware of the gap between China and the West in science and technology, and actively contributed to change the situation. For example, during the Opium Wars, the Hong merchants took the lead role in donating money to introduce advanced Western technology and upgrade the equipment of Guangdong Navy; Pan Shirong, a Hong merchant, made efforts to imitate the most advanced steamships in the world at that time; Pan Shicheng hired American navy officers with high salary to copy the mines and created the earliest modern Chinese mine, which was called “ship-attacking mine”; Zheng Chongqian, the owner of Huilong Hong, actively promoted

May / 2021.5

19

星期三

Wednesday

辛丑牛年 癸巳月

四月初八

the injection of vaccine against smallpox, and was the first Chinese to promote the method; the Wu family had contributed a lot to the introduction of Western medical science into China. The first hospital to promote vaccination, hospital of vaccination against smallpox, and the Ophthalmic Hospital set up by Peter Parker, had received strong support from the Wu family and other Hong merchant families. It is justifiable to say that the Hong merchants were the first practitioners who insisted on "learning from the foreigners in order to control them", and their activities were 20 years earlier than the official Westernization Movement.

19世纪初期的画作：广州十三行。作为鸦片战争前中国对西方进行海上贸易的唯一关口，这里诞生了中国第一批"开眼看世界"的人。

A painting of the Thirteen Hongs in Canton, early 19th century. Canton was the only Chinese port that conducted maritime trade with the western countries, and here came the first group of Chinese people who "opened their eyes to see the world".

May / 2021.5

20

星期四

Thursday

辛丑牛年 癸巳月

四月初九

行商美庭园

The Hong Merchants' Gardens

“行商庭园”是由岭南著名建筑大师莫伯治率先提出的概念，指清代由十三行行商们所修建的私家园林。行商庭园集中分布于广州城西、河南、花地一带(今海珠、荔湾),除作为行商生活起居之所外,也是与洋商联谊的重要场所,在广州的对外事务上发挥着极重要的作用。

清政府规定外国人不许随意在广州走动，而行商庭园是他们能活动的有限场所之一。这些精美的岭南园林及造园思想令他们印象深刻，是当时外销画中常见的主题，并对 18 世纪欧洲宫廷的审美意趣产生极大影响。法国、英国、德国的宫苑园林里，都出现了亭、榭、塔、桥和假山曲径等中国式元素，仅巴黎就有 20 多处中国式风景园林。这股对于中国意象的追捧和仰慕对一种新的艺术风格——洛可可的形成产生了重要的影响。

“The Hong Merchants' Gardens” was a concept first proposed by Mo Bozhi, a famous architectural master in Lingnan region. It refers to the private gardens built by the Hong merchants in the Qing dynasty. The gardens were mainly situated in the western part of Canton, the southern bank of the Pearl River and Huadi area (the land of flowers, now in Haizhu and Liwan district). They were not only places of the Hong merchants' daily life, but also important venues to receive foreign merchants, and played very important roles in the foreign affairs in Canton.

The Qing government stipulated that foreigners were not allowed to travel freely in Canton, and the Hong merchants' gardens were one of the few places open to them. They were deeply impressed by these exquisite Lingnan gardens and architectural ideas, which were common themes in the export paintings at that time, having a great impact on the aesthetic interest of the European

Grain Buds

小满

May / 2021.5

21

星期五

Friday

辛丑牛年 癸巳月

四月初十

courts in the 18th century. In France, England and Germany, there are Chinese architectural elements such as pavilions, houses, towers, bridges rockeries and winding paths in the gardens. There are more than 20 Chinese landscape gardens in Paris alone. The pursuit and admiration for Chinese imagery had a significant influence on a new artistic form, the Rococo style.

19世纪末象牙透雕扇骨蝴蝶寿字纹折扇。

Folding ivory fan with pierced patterns of butterflies and characters of longevity, 19th century.

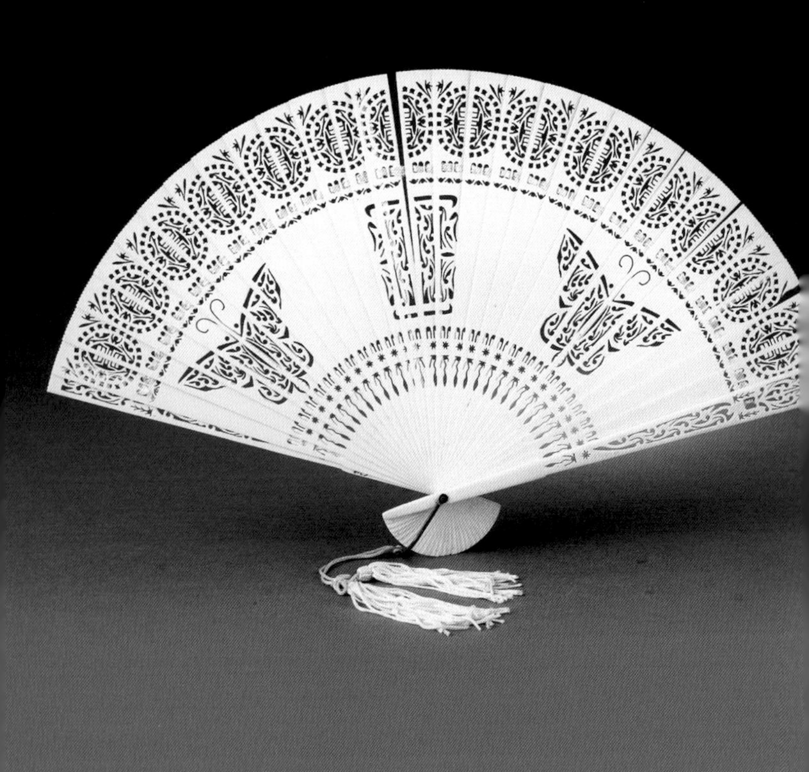

May / 2021.5

22

星期六

Saturday

辛丑牛年 癸巳月

四月十一

清末玳瑁扇骨彩绘花鸟纹羽毛折扇。

Folding feather fan made of hawksbill shell with polychrome patterns of flowers and birds, late Qing dynasty.

May / 2021.5

23

星期日

Sunday

辛丑牛年 癸巳月

四月十二

伍家有花园

The Wu Family's Garden

在当时广州众多的行商庭园里，最被西方人频繁提到的是“浩官花园”，也就是怡和行伍氏家族的伍家花园。19 世纪来华的西方人都以能造访伍家花园为荣。

伍家花园主要包括河南伍园和花地馥荫园两座园子。河南伍园位于珠江南岸，建于嘉庆八年（1803 年）是伍氏家族根基所在，家宅、宗祠皆建于此，还有一座精美的园林——万松园，主要建筑包括粤雅堂、宝纶楼、听钟楼、揖山楼、听涛楼、延晖楼、枕流室等，是一个祠、邸、园及原有景物融汇一体的宏大园林景观建筑群。

馥荫园原本是同文行潘氏的东园，被伍氏接手后改为馥荫园。花地自古花场云集，是当时西方博物学家采集广东植物花卉样本的主要场所。位于花地河入口东南侧的馥荫园自然也以花木种植闻名，是伍氏避暑消夏和接待外国商人的地方。

Among the numerous Hong merchants' gardens in Canton at that time, the one most frequently mentioned by Westerners was “Howqua's garden”, the garden of the Wu family who owned Yihe Hong. Westerners who came to China in the 19th century would feel proud if they could visit Howqua's garden.

The Wu family owned two gardens: the first one was Wu's Henan Garden, and the second was Fuyin Garden. Wu's Henan Garden was located on the southern bank of the Pearl River in Canton. It was built in the eighth year of Emperor Jiaqing's reign (1803), as the foundation of the Wu family. Inside, there were houses, an ancestral hall and a beautiful garden–named Ten Thousand Pine Trees Garden. The main

May / 2021.5

24

星期一

Monday

辛丑牛年 癸巳月

四月十三

buildings include Yueya Hall (Hall of Elegance in Canton), Baolun Building, Tingzhong Building, Yishan Building, Tingtao Building, Yanhui Building, Zhenliu Room, etc. it was an architectural complex with temples, residences, and gardens, perfectly matched with the local scenery.

Fuyin Garden, originally as the Pan family's East Garden, was taken over and converted by the Wu family. Huadi, the location of the garden had been a place with numerous flower plantations since ancient times, and became the main location for Western naturalists to collect samples of Cantonese plants and flowers. Fuyin Garden, located in the southeastern side of the estuary of Huadi River, famous for its flowers and trees, was a place for Wu family to spend summer and receive foreign merchants.

外销画中的伍家花园。（左图）（原件藏于广州十三行博物馆）

The Wu family's garden in the export paintings. (left) (The original piece is collected in Guangzhou Thirteen Hongs Museum.)

19世纪纸本水彩园景仕女图。（右图）（原件藏于广州十三行博物馆）

Ladies in the garden, watercolour on paper, 19th century. (right) (The original piece is collected in Guangzhou Thirteen Hongs Museum.)

May / 2021.5

25

星期二

Tuesday

辛丑牛年 癸巳月

四月十四

潘园著风雅

The Elegant Pan Garden

1776 年，来自福建的同文行行商潘振承买下今广州海珠区漱珠涌一带 20 公顷河洲地，在这里开基立祠，以福建老家的名字命之为龙溪乡，并修建了漱珠、环珠和跃龙三桥，由此拉开了潘家花园长达百年的建造序幕。

整个潘家花园系一大型的私家园林别墅群，以一个巨大的方形池塘为中心，由潘家子孙沿塘先后建造庭院、别墅、书斋、花园等各种建筑，如潘有为所建的六松园、南雪巢、橘绿橙黄山馆、看篆楼；潘有度所建的漱石山房、义松堂、南墅；还有潘正炜所建的清华池馆、听飒楼等，规模宏大，雍容大气，彼此借景又各有特色。时人有诗云："芳园风日倍精神，屈曲高低妙出新。傍水楼台都入画，引雏莺燕故随人。"漱珠涌也成为当时最为繁华风雅的地方。

In 1776, Pan Zhencheng, a Hong merchant from Fujian Province, bought a batture land of 20 hectares in the area of today's Shuzhu Creek in Haizhu district of Canton, on which he laid the foundation, built a temple and three bridges called Shuzhu, Huanzhu and Yuelong. The Pan Garden was named after his hometown in Fujian as Longxi Town and thus the construction lasting a hundred year was started.

The whole Pan Garden was a large-scale complex of private villas and gardens, with a huge square pond in the center. The Pan family's descendants had successively built courtyards, villas, libraries, gardens and other buildings surrounding the pond, such as the Liusong Garden, Nanxuechao, Julvchenghuang Pavilion and Kanzhuan Building built by Pan Youwei, Shushishan House,Yisong Hall, Nan Villa built by Pan Youdu, Hall of Qinghua Pool, and Tingfeng Building built by

May / 2021.5

26

星期三

Wednesday

辛丑牛年 癸巳月

四月十五

Pan Zhengwei, etc. The buildings were grand in size, luxurious and elegant, with different characteristics forming contrast among one another. At that time, there was a poem saying: "the wind and the sun in the garden make you feel high-spirited; the ups and downs of the buildings are wonderful and novel. The buildings near the water are apt for a painting, making the squabs and swallows follow the passers-by." Shuzhu Creek was also the most prosperous and elegant place at that time.

潘有度家的庭园。（图片采自《大英图书馆特藏中国清代外销画精华（第三卷）》）

Pan Youdu's garden. (The picture is taken from *Chinese Export Paintings of the Qing Period in The British Library. Vol III.)*

May / 2021.5

27

星期四

Thursday

辛丑牛年 癸巳月

四月十六

仙馆誉岭南

The Garden of the Immortals

清道光十年（1830 年），潘振承的族曾孙潘仕成买下西关荔枝湾方圆数百亩地，开始建造号称广州私园之冠的海山仙馆。海山仙馆所在的位置是南汉王朝的御花园昌华苑旧址，西、南面有珠江与白鹅潭，东边是西关民居，北边是田野和山岗。馆内开凿了百亩大湖，并堆土成山，遍植花草果木，不以华丽取胜，而是将文化传承、地域特色与田园景观三者融为一体，情景互衬，天人合一，体现了岭南园林艺术最高超的境界与神髓——素雅及精致，时人誉为“妙有江南泂水意，却添湾上荔枝多”。

由于潘仕成亦官亦商的身份，海山仙馆成为重要的外事活动场所，不仅接待过美国作家亨特、法国摄影师于勒·埃及尔、英国著名的摄影师和作家约翰·汤逊，甚至 1846 年美国首任驻华公使义华业向耆英递交美国总统致清廷国书的仪式都在这里举行。

In the tenth year of Emperor Daoguang's reign (1830), Pan Shicheng, the great grandson of Pan Zhencheng, bought a land of hundreds of acres near the Lizhi Creek in Xiguan area, and began to build Haishanxianguan (Garden of the Immortals on the Sea and Mountain), known as the best private garden in Canton. The location was the former site of Changhua Garden of the Southern Han Kingdom. To its west and south, there were the Pearl River and White Swan Pond; to its east there were civilian residences in Xiguan area, and in the north, fields and hills. In the garden, there was a great lake of one hundred acre, and piled up rockeries; flowers, plants, fruit trees were everywhere. Instead of being characterized by luxuriousness, the design integrated the cultural heritage, regional characteristics and rural landscapes into one piece. With the scenes interlinking with one other, and the connotation

May / 2021.5

28

星期五

Friday

辛丑牛年 癸巳月

四月十七

of a union of nature and human, it had represented the highest realm and spiritual marrow of Lingnan landscape art——to be elegant and refined. A poem described it as follows: "the watery scenes are beautiful in Jiangnan (the southern region of the Yangtze River), and many lychees are ripe on the trees beside the creek".

Pan Shicheng was an official and meanwhile a Hong merchant. His garden became an important place for activities of foreign affairs, not only having received William C. Hunter, an American writer, Jules Itier, a French photographer, John Thomson, a famous British photographer and writer, but also having held the ceremony when in 1846, Alexande Hill Everett, the first American envoy to China presented his credentials to deliver the letter from the president of the United States to the Qing court through the official Qi Ying.

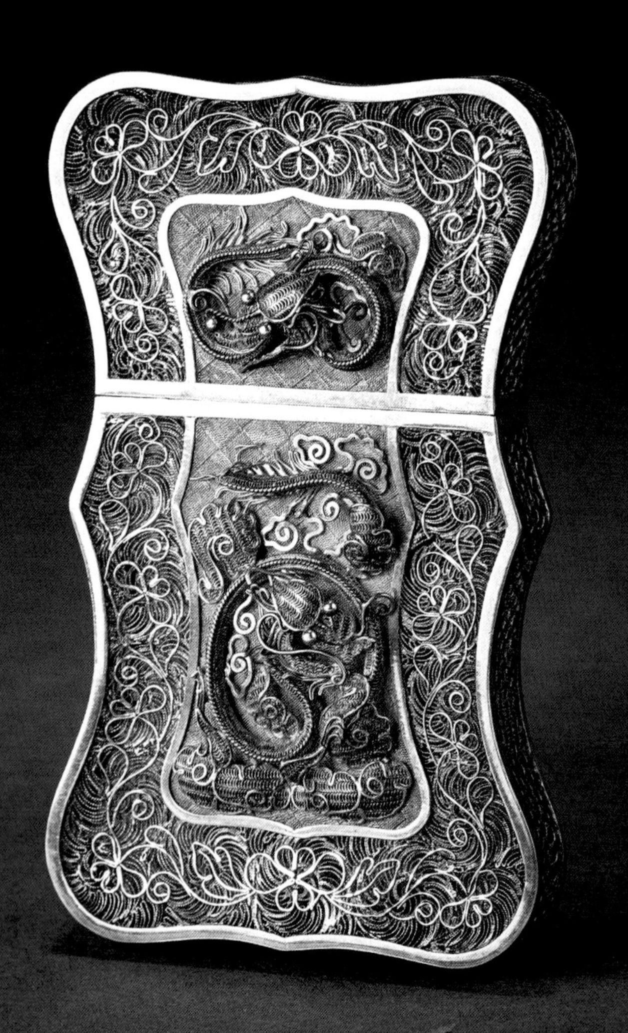

May / 2021.5

29

星期六

Saturday

辛丑牛年 癸巳月

四月十八

清末银花卉龙纹名片盒。

Silver business card box with patterns of flowers and dragon, late Qing dynasty.

19世纪银缧丝镀金烧珐琅彩花卉船舶图双龙戏珠盒。

Gilded silver filigree enamel covered box with patterns of flowers, and dragons playing the pearl, late 19th century.

May / 2021.5

30

星期日

Sunday

辛丑牛年 癸巳月

四月十九

无边风月散

All the Fortune Had Gone

在整个广州城市发展史里，行商庭园可谓惊鸿一瞥，它在短短百年里迅速兴起又迅速衰败，只有零星故迹留存至今。

第一次鸦片战争之后，十三行渐渐败落乃至瓦解。作为十三行的衍生物，珠江西岸众多的行商庭园也随之萧索。而更主要的原因是清政府长期以来对十三行行商各种严苛的规定、勒逼、盘剥，让行商们稍不注意就获罪流放、家财尽散。如海山仙馆因潘氏获罪而被朝廷抄没拍卖，随后被支离拆卖而消失，只存在了40余年；潘家花园因潘家中落被没收充公；伍家花园、潘长耀花园则或因业务收缩，或因商行倒闭而逐渐败落。此后，随着清末民初广州城市人口激增，河南地区被大规模开放，越来越多的民居侵占园地，这些精美的园林最终消失在民宅街巷之间。

In the whole history of Canton's urban development, the Hong merchants' gardens could be deemed as an astonishing phenomenon, which prospered and then declined rapidly in just a hundred year. Since then only a few parts of them have remained.

After the First Opium War, the merchant houses gradually declined and eventually closed down. As derivatives of the Canton trade, many Hong merchants' gardens on the western bank of the Pearl River became wasted. But the more important reason was that the Qing government had long imposed various strict regulations, extortion and exploitation on the Hong merchants. If they were not careful enough, they might end up with exile or squandering all their fortune. For example, Haishanxianguan, (Garden of Immortals on the Sea and Mountain) was confiscated and auctioned by the court because of its owner Pan had been

May / 2021.5

31

星期一

Monday

辛丑牛年 癸巳月

四月二十

penalized. Having existed for more than 40 years, the garden disappeared because of being dismantled and sold. The Pan family's garden was confiscated because of their fall, while the gardens of the Wu family and Pan Changyao also gradually declined because of their contraction in business or bankruptcy. Since then, with the rapid increase of urban population in Canton at the end of the Qing dynasty and the beginning of the Republic of China, the area of the southern bank of the Pearl River had been largely developed, and more and more residential buildings had occupied the locations of the gardens. These beautiful gardens eventually disappeared and were replaced by the residential streets.

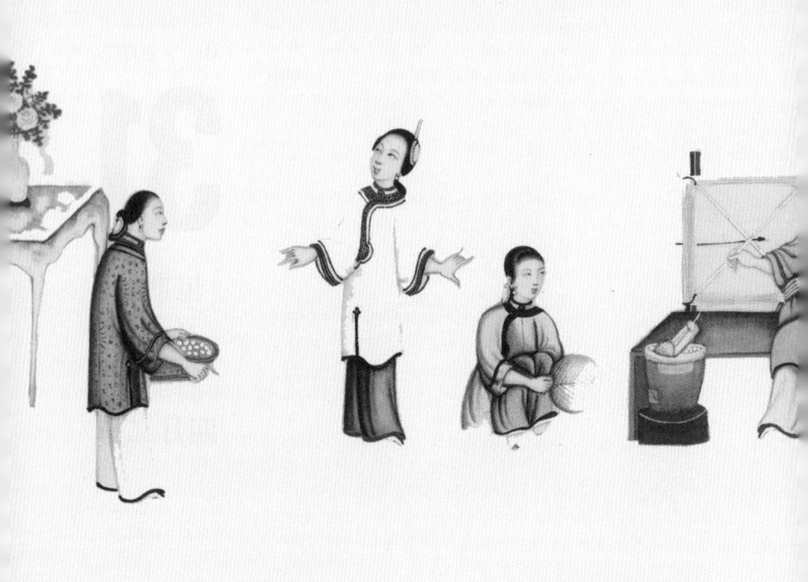

June

一	二	三	四	五	六	日
	1 国际儿童节	2 廿二	3 廿三	4 廿四	5 芒种	6 廿六
7 廿七	8 廿八	9 廿九	10 五月	11 初二	12 初三	13 初四
14 端午节	15 初六	16 初七	17 初八	18 初九	19 初十	20 十一
21 夏至	22 十三	23 十四	24 十五	25 十六	26 十七	27 十八
28 十九	29 二十	30 廿一				

法国摄影师于勒·埃及尔拍摄的海山仙馆一角。(图片来源：FOTOE 图片库)

A small part of Haishanxianguan, by Jules Itier, a French photographer. (The picture is bought from the photo gallery FOTOE.)

June / 2021.6

1

星期二

Tuesday

辛丑牛年 癸巳月

四月廿一

国际儿童节

International Children's Day

创造“土”英语

Cantonese English

为了便于与外国商人交流做生意，外语成为十三行商人的必备技能。大约从乾隆十五年（1750 年）开始，广州出现了所谓的“广东英语”，指行商用广州话注音，带有浓重广州话口音的英语。据司徒尚纪教授统计，至 1949 年，用广州话标注的“广东英语”单词有 160 多个，几乎都是围绕着贸易展开的。例如：Have you bought some tea from him?（你是否曾经向他买茶？）广东话标注为：“哈父天砵心哋父林谦？”

“广东英语”虽然发音不标准，但是在谈生意时很管用，十三行行商甚至承担起政府的翻译工作。英国贡使入京见嘉庆皇帝的时候，嘉庆皇帝宣召两名熟悉英语的行商作为翻译；两广总督收到外国大班的公文，也是由行商翻译成中文的。“广东英语”后来随行商到上海经商而流行于洋泾浜，演变成“洋泾浜英语”。

In order to trade and communicate with foreign businessmen, learning foreign languages had become a necessary skill for the Hong merchants. From around the fifteenth year of Emperor Qianlong's reign (1750), the so-called “Cantonese English” appeared, and it referred to the English with a strong Cantonese accent. The Hong merchants used local Cantonese to represent the pronunciation of English. According to the statistics of Professor Situ Shangji, by 1949, there had been more than 160 “Cantonese English” words marked with Cantonese dialect, most of which were related to trade. For example, the sentence “have you bought some tea from him?” would be represented as “哈父天砵心哋父林谦？” in Cantonese.

The pronunciation of “Cantonese English” was not standardized, but it was

June / 2021.6

2

星期三

Wednesday

辛丑牛年 癸巳月

四月廿二

very useful in business negotiations, and the Hong merchants even undertook the translation for the government. When the British tribute envoys came to Beijing to visit Emperor Jiaqing, two Hong merchants, who were familiar with English, served as translators. When the governor of Guangdong and Guangxi provinces received the official letters from the foreign comprador, the Hong merchants translated the letter into Chinese. "Cantonese English" later became popular when the Hong merchants traded in Shanghai, and evolved into "Pidgin English".

June / 2021.6

3

星期四

Thursday

辛丑牛年 癸巳月

四月廿三

英国画家乔治·钱纳利作品：郭雷枢医生诊所。和皮尔逊医生一样，郭雷枢也是东印度公司雇用的专业医生，在来广州做传教医生前已经做过五年商船外科医生。（图片采自《东西汇流》）

Dr. Thomas Richardson Colledge's clinic, painted by George Chinnery, an English painter. Dr. Thomas Richardson Colledge, like Dr. Alexander Pearson, was a professional doctor employed by the English East India Company. Before he came to Canton as a missionary, he had already worked for 5 years as a surgeon on the merchant ships. (The picture is taken from *East Meets West*.)

做功牛痘法

Vaccination against Smallpox

1796年，英国医生琴纳发明了接种牛痘预防天花的方法。19世纪初，牛痘接种法经“海上丝绸之路”传到澳门，由英国东印度公司的皮尔逊医生开办了诊所，在嘉庆十一年（1806年）广东地区爆发天花病时，以接种方式挽救了数千人性命。皮尔逊收了四个中国人当助手，其中一位叫邱熺的商人学得最好，他把牛痘接种法带到了广州，得到了十三行行商的大力支持，合资向大众免费推广牛痘接种法。自嘉庆十五年至道光二十年（1810—1840年），广东接种牛痘者达到30万人次之多。道光八年（1828年），行商潘仕成把种牛痘局开到了北京宣武门外上斜街南海邑会馆。此后50年，牛痘接种法依次传至湖南、湖北、江苏、安徽、江西、浙江、四川、天津、山东、广西等地乃至全国，对中国预防天花做出了很大贡献。

In 1796, Edward Jenner, a British doctor, invented the vaccine against smallpox, which was spread to Macao through the “Maritime Silk Route” at the beginning of the 19th century. Alexander Pearson, a doctor who worked for the English East India Company, opened a clinic and saved thousands of lives with vaccination in the eleventh year of Emperor Jiaqing’s reign (1806), when smallpox broke out in Guangdong province. Alexander Pearson received four Chinese as assistants, and there was a merchant called Qiu Xi who learned the best. The introduction of vaccination into Canton was strongly supported by the Hong merchants. With the capital from the Hong merchants, Qiu promoted vaccination to the public free of charge. From the fifteenth year of Emperor Jiaqing’s reign to the twentieth year of Emperor Daoguang’s reign (1810-1840), the number of Cantonese people who had received the vaccination was up to 300,000. In the eighth year of Emperor

June / 2021.6

4

星期五

Friday

辛丑牛年 癸巳月

四月廿四

Daoguang's reign (1828), the Hong merchant Pan Shicheng opened the vaccination clinic in Nanhaiyi Hall, whose location was outside of the Xuanwu Gate in Shangxie Street, Beijing. In the following 50 years, vaccination was successively introduced to other provinces such as Hunan, Hubei, Jiangsu, Anhui, Jiangxi, Zhejiang, Sichuan, Tianjin, Shandong, Guangxi and even the whole country. This was a great contribution to curb smallpox in China.

19世纪贝雕镂空宗教人物图挂件。

Carred shell with pierced patterns of religions stories, 19th century.

芒种

Grain in Ear

June / 2021.6

5

星期六

Saturday

辛丑牛年 甲午月

四月廿五

清晚期描金绘贝摆件。

Shell with gold-painted patterns, late Qing dynasty.

June / 2021.6

6

星期日

Sunday

辛丑牛年 甲午月

四月廿六

眼科开先河

The First Hospital of Ophthalmology

十三行商区还诞生了中国最早的眼科专科医院。

1834 年，美国传教医师伯驾来到广州，得到怡和行等行行商的支持，在新豆栏街丰泰行 3 号租借一幢商人的房子，于 11 月 4 日开设“眼科医局”（又称“新豆栏医局”）。这个医院的规模不算小，设有接待室、诊断室、配药室、手术室和观察室，能容 200 个病人候诊。伯驾在这里成功施行了第一例白内障手术。由于求医的人太多，怡和行又把丰泰行 7 号楼房免费借给伯驾扩充医局分院，起名为“博爱医院”（即后来博济医院前身）。1865 年，博爱医院迁至长堤现址，更名“博济医院”；1866 年，博济医院增设“博济医学堂”，后改名为“南华医学院”，并于 1879 年开始招收女生。1880 年，博济医院创办《中华医报》，揭开了中国现代医学杂志史的第一页。

The earliest ophthalmic hospital in China was set up in the trading region.

In 1834, Peter Parker, an American physician and missionary, came to Canton and gained support from Yihe Hong and other Hong merchants. He rented a room in Fengtai Hong at No. 3, Hog Lane. On November 4th, he opened the “Medical Bureau of Ophthalmology”, also known as “Medical Bureau in Hog Lane”. This hospital, with reception room, diagnosis room, dispensing room, operating room and observation room inside, was large enough to hold 200 patients waiting for treatment. Peter Parker successfully performed the first cataract operation here. Due to the large number of people who sought for medical treatment, Yihe Hong rented the No. 7 building of Fengtai Hong to Peter Parker for free to expand the parallel section of the hospital, which was called

June / 2021.6

7

星期一

Monday

辛丑牛年 甲午月

四月廿七

“Bo’ai Hospital” (Hospital of Fraternity), which is the predecessor of Boji Hospital (Hospital of Succour) in the later period. In 1865, Bo’ai Hospital was moved to the current site of Changdi Road, the bank of the Pearl River, and re-named as “Boji Hospital”; in 1866, Boji Hospital established “Boji medical school”, which was later re-named as “Nanhua Medical College”, and began to recruit girl students from 1879. In 1880, Boji Hospital founded the *Chinese Medical Journal*, which was the beginning of the history of modern Chinese medical journals.

June / 2021.6

8

星期二

Tuesday

辛丑牛年 甲午月

四月廿八

美国传教士兼医生彼得·伯驾博士和他的助手，在广州诊所为患者做眼科手术。关乔昌绘于 1839 年。（图片来源：FOTOE 图片库）

Peter Parker, an American doctor and missionary, and his assistant were treating a patient who had eye disease. A painting by Guan Qiaochang, Lamqua, 1839. (The picture is bought from the photo gallery FOTOE.)

倾力兴文教

Dedication to Cultural Education

大约从乾隆二十年（1755 年）开始，在行商捐资办学的影响下，广东地区掀起了一股兴办书院热潮。到了道光二十年（1840 年）前后，仅广州就有包括越华、文澜等知名书院在内的书院 30 间。广州书院除了教授传统文学、音韵、训诂等传统课程外，还仿照近代西方国家学校增设了数学、天文、地理、历法等自然科学知识的课程，培养出很多广东近代革新派的思想家，为中国社会近代化进程做出了不可磨灭的贡献。广东也是最积极响应 1909 年“废科举，兴学堂”浪潮的地区，不仅将书院改为学堂，还兴办了一批新式学堂，其中包括诸如培正、真光、培英、东山、金山等名校，客观上推动了全国的教育改革。

Since the twentieth year of Emperor Qianlong's reign (1755) or so, under the influence of the Hong merchants who donated to establish schools, there had been an upsurge of establishing academies in Guangdong province. By the twentieth year of Emperor Daoguang's reign (1840), Canton alone had 30 academies, including Yuehua, Wenlan and other famous academies. In addition to teaching traditional courses such as traditional literature, phonology and exegesis, the Cantonese academies also learned from the Western schools to set up natural and scientific courses like mathematics, astronomy, geography and calendar, nurturing many thinkers of the modern innovation in Guangdong area, and making indelible contributions to the process of modernization of Chinese society. Guangdong also actively responded to the wave of “abolishing the imperial examination and re-vitalizing the schools”, a notion put up in 1909, not only changing the academies into schools, but also setting up a number of new schools for modern education, including the famous ones such as Peizheng, Zhenguang, Peiying, Dongshan and Jinshan, which objectively promoted the national educational reform.

June / 2021.6

9

星期三

Wednesday

辛丑牛年 甲午月

四月廿九

20世纪初期，广州夏葛女医学堂的两位女毕业生。（图片来源：FOTOE图片库）

Two female graduates from Xiage Girls' Medical School in Canton, early 20th century. (The picture is bought from the photo gallery FOTOE.)

June / 2021.6

10

星期四

Thursday

辛丑牛年 甲午月

五月初一

商行重棉纺

The Cotton Textile

清代广州出产的棉花，以轻暖闻名，号称“广花”，织出的棉布较英国棉布物美价廉，是十三行对外出口的大宗商品之一。旺盛的对外贸易需求带动了广州纺织业的发展，西关大片区域被辟为街道，设立厂房，成为了广州的纺织业基地，涌现出一批与纺织有关的地名，如锦华大街、经纶大街、麻纱巷等。纺织业的发展带动了印染、晒、浆缎、机具、制衣、制帽、鞋、袜、绒线等相关行业的兴起，西关第六甫、第七甫、第八甫至长寿里、青紫坊、芦排巷一带成为近代工业区——机房区。由纺织业发家的富商们纷纷在靠近机房区的西关角一带兴建住宅，使这一片地区发展成为高档住宅区——西关大屋，继而由此诞生出广州近代女性人文标志——西关小姐。

The cotton produced in Canton in the Qing dynasty was famous for its lightness and warmth, known as the "Cantonese cotton". Compared with the British cotton, the Cantonese cotton was cheaper and better in quality. It was one of the bulk stock commodities exported to the outside world through Canton trade. The vital demand of foreign trade led to the development of Cantonese textile industry. Large part of the Xiguan area region was turned into streets where factories had been set up. As a textile center in Canton, some places would be named after the terms related to textile, such as Jinhua (colorful satin) Street, Jinglun (silk thread) Street, and Masha (cambric gauze) Lane. The development of textile industry also led to the rise of other related industries such as printing, sun-drying, dyeing, machine-building, clothing, hat-making, shoe-making, sock-making, wool-producing, etc. Some places in Xiguan area, such as Diliufu, Diqifu, Dibafu, Changshouli, Qingzifang, and Lupaixiang became the modern industrial zone——the region of factories.

June / 2021.6

11

星期五

Friday

辛丑牛年 甲午月

五月初二

The rich traders who made their first fortune from textile industry built houses in Xiguan corner near the factories, and this area developed into a high-end residential area——the grand houses in Xiguan thus appeared. This was also the birth of Ladies in Xiguan area, the cultural symbol of modern women in Canton.

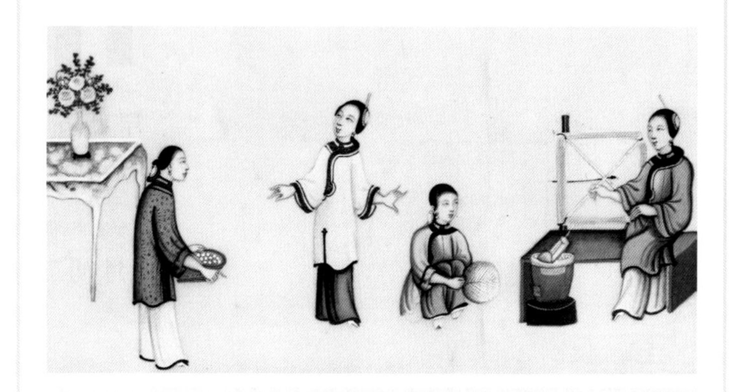

19世纪通草水彩纺织图。

Textile process, watercolour on pith paper, 19th century.

June / 2021.6

12

星期六

Saturday

辛丑牛年 甲午月

五月初三

19世纪末通草水彩市井人物图。

People in the market, watercolour on pith paper, late 19th century.

June / 2021.6

13

星期日

Sunday

辛丑牛年 甲午月

五月初四

西关有大屋

Grand Houses in Xiguan Area

西关大屋兴起于清末，延续到民国，其形制按时代的早晚分为岭南传统建筑、中西结合的宅子、纯西式洋楼和红砖小楼房，反映了不同时期广州上流社会的审美情趣和流行风尚，具有浓郁的时代特征。

西关大屋的特征是平面呈狭长的方形，面阔小而进深大，很好地适应了岭南地区夏季高温、多雨、潮湿的气候特征。狭长的房屋布局能增强空气对流，再配以院子、天井、敞口厅、青云巷、天窗、侧窗，以及可以活动的屏风、满洲窗等结构，将穿堂风聚拢增强，就会令房子冬暖夏凉，舒适宜居。西关大屋主体结构为平面三间两廊，主要厅堂居于纵向中轴线，两侧是对称的偏厅、小院、书房、客房、卧房、厨房及佣人房。最外层为两条贯通的巷道，有通风、防火、排水、采光、晒晾、交通、栽花木等多种功用。

The grand houses in Xiguan area first appeared in the late Qing dynasty and continued to be built until the Republic of China. According to different periods of time, their shapes are divided into Lingnan traditional architectural style, Sino-Western style, Western style and red-brick style of small villa, reflecting the aesthetic taste and popular fashion of the upper class in Canton and strong features of different eras.

A grand house in Xiguan, characterized by a long and narrow rectangular shape, with a cramped width and profound depth, could be well adapted to the rainy, humid and hot weather of summer in Lingnan region. The long and narrow layout of the house enhances the air ventilation, while the courtyard, patio, open hall, a small lane that separates the house form another, skylights, side windows, movable screens, Manchurian windows and other structures are apt for ventilation,

June / 2021.6

14

星期一

Monday

辛丑牛年 甲午月

五月初五

端午节

Dragon Boat Festival

making the house warm in winter while cool in summer, and comfortable to live in. The main structure of a house is made up of three halls and two corridors. The main hall is located in the longitudinal axis, with a pair of symmetrical partial halls, terraces, study rooms, guest rooms, bedrooms, kitchens and servant rooms on both sides. The outer layer is surrounded by two connected tunnels which have many functions such as ventilation, fire prevention, drainage, absorbing daylight, cloth-drying, transporting, and planting flowers and trees.

June / 2021.6

15

星期二

Tuesday

辛丑牛年 甲午月

五月初六

西关大屋的主厅堂。（图片来源：FOTOE 图片库）

The main hall of a grand house in Xiguan area. (The picture is bought from the photo gallery FOTOE.)

大屋三件宝 Ⅰ

Three Precious Objects of the Grand House Ⅰ

提到西关大屋，老广州们通常都会用一句俗语来概括：“西关大屋三件宝，青砖石脚三重门。”

青砖和石脚的作用在于防潮。青砖产自东莞、清远一带，砌于外墙，以色泽青碧如玉者为上佳。好的青砖墙不仅要求青砖质地坚硬，外形规整，对砌者的手艺也有很高要求，讲究的青砖墙砖缝之间连刀片都插不进去，可历经数百年而不磨蚀。石脚用打磨平滑的花岗石拼砌而成，类似台基，距地面高度从 70 至 100 厘米不等。石脚讲求接缝越少越好，因此多以大块整石构建，十分耗工耗银。渐渐的，青砖和石脚还附加了“炫富”功能，哪座大屋青砖的颜色越青、接缝越细密，砌石脚的石块越大越平滑、接缝数量越少、石脚高度越高，就表示屋主的财富越多。

When it comes to the grand houses in Xiguan, the local people in Canton usually use a common saying to summarize: “there are three precious objects in a grand house of Xiguan: the blue bricks, stone steps and a triple gate.”

The purpose of using blue bricks and stone steps is to prevent moisture. blue bricks are produced in Dongguan and Qingyuan. They are used to build the outer wall. The color of blue bricks is as good as jade. A brick wall of good quality is hard in texture with a regular and neat appearance, which is a high requirement for the the bricklayers' craftsmanship. Even a blade cannot insert into the joints between the bricks of an exquisite wall, which can last hundreds of years without abrasion. The stone step is made of polished granite, similar to the platform base, with a height from 70cm to 100cm above the ground. The lesser the joints in a stone step, the better the quality is. Therefore, the stone steps are mostly made of large

June / 2021.6

16

星期三

Wednesday

辛丑牛年 甲午月

五月初七

stone blocks, which are very expensive and the processing must take a lot of work. Gradually, the blue brick steps also bore the function of "showing off". The greener the bricks of the grand house were with closer joints, the smoother and higher the stone steps were with fewer joints, the wealthier the owner might be.

June / 2021.6

17

星期四

Thursday

辛丑牛年 甲午月

五月初八

西关大屋的正门，由矮脚吊扇门、趟栊门、硬木大门组成。（图片来源：FOTOE 图片库）

The front door of a grand house in Xiguan is made up of a hanging door with a low foot, a sliding door and a big door made of hard wood. (The picture is bought from the photo gallery FOTOE.)

大屋三件宝 Ⅱ

Three Precious Objects of the Grand House Ⅱ

三重门是岭南人为适应气候环境而创造的一个奇妙发明，集安全、通风、私密性于一体。临街第一重门叫“矮脚吊扇门”，简称脚门，外形像四折屏风。脚门上半部分雕成精致通透的花纹，题材丰富多样，有吉祥花鸟图、菱形网格图、平安花瓶图等。第二重门即趟栊门，为岭南特有建筑构件，因独特的横向开合方式而得名。趟栊门像打横放置的栅栏，有数条粗大坚硬的圆木镶于两根竖板上，竖板下方安有滑轮，可以沿着镶于门槛上的铁轨左右拉动。它的门锁为木制的顶闩，隐藏在另一侧的大门后，一旦有外人硬闯，屋内人只需将闩轻轻按下，即可顶死趟栊门令其不能开合。第三重是西关大屋真正意义上的大门，分左、右两扇，或施朱或染墨，再刷以桐油清漆，配上黄澄澄的铜环拉手，显得厚实端庄。

The triple gate (a gate with three layers) is a wonderful invention created by Lingnan residents to adapt to the environment and climate, integrating safety, ventilation and privacy. The first layer facing the street is called “hanging door with a low foot”, also briefly called as the “footed door”, and looks like a four-fold screen. The upper part of the footed door is carved with delicate and transparent patterns, with various themes, including auspicious flowers and birds, diamond-shaped grid, vases symbolizing safety, etc. The second gate is the sliding gate, which is a unique architectural component of Lingnan. It is named after its unique transverse mode of opening and closing. The sliding door is like a fence placed horizontally. A few thick and hard logs would be inlaid on two vertical plates, under which there are pulleys. The door can be slid left and right along the rail on the threshold. The door lock is a wooden bolt, hidden on the back of the first gate. If someone violently intrudes

June / 2021.6

18

星期五

Friday

辛丑牛年 甲午月

五月初九

the house, people inside just need to press the bolt gently, and the door will remain closed. The third layer is the gate of the grand house in a real sense, which is divided into left and right halves, painted red or black, varnished with tung oil, and installed with yellow copper ring handles. The gate looks solid and dignified.

清代通草水彩吸鸦片图。

Smoking opium, watercolour on pith paper, Qing dynasty.

June / 2021.6

19

星期六

Saturday

辛丑牛年 甲午月

五月初十

19世纪通草水彩婚俗图。

The wedding customs, watercolour on pith paper, 19th century.

19世纪通草水彩戏曲故事图。

The opera scene, watercolour on pith paper, 19th century.

June / 2021.6

20

星期日

Sunday

辛丑牛年 甲午月

五月十一

华夷满洲窗

Manchurian Windows

西关大屋最引人注目的莫过于精美的满洲窗。满洲窗由传统的木框架镶嵌套色玻璃蚀刻画而成，最早兴于乾隆时期，是驻防广州的满族八旗兵自北方带来的满族窗户样式，后来渐渐流行于西关地区，在富裕阶层中蔚为风尚。随着十三行对外贸易的繁荣，西方的彩色玻璃和套色玻璃蚀刻画技术传入广州，与原有的满洲窗样式结合起来，形成了彩色满洲窗。

套色玻璃蚀刻画本身也是一种中西合璧的工艺，它采用蚀刻、磨刻或喷沙脱色等西式技术对进口玻璃材料进行处理，而图案的内容和绘制手法则遵循中国传统文化。玻璃蚀刻画有红、黄、蓝、绿、紫、金等色，通过不同形状的窗框与窗格，将色彩加以分割搭配，使窗户显得明艳通透、雍容典雅，是岭南建筑艺术的瑰宝，也是广府文化的重要符号之一。

The most attractive parts of a grand house in Xiguan are the exquisite Manchurian windows. Manchurian windows are made of etched colorful glass paintings inlaid in traditional wood frames. First flourishing in the Emperor Qianlong's reign, the style originated from the window patterns brought by the Manchurian garrison soldiers who came from the northern area and carried the garrison duty in Canton. Later, the style gradually became popular in Xiguan area as a fashion among the rich. With the prosperity of foreign trade, the Western colourful reverse glass paintings and chromatic glass etching were introduced into Canton, and combined with the original style of Manchurian windows. Thus the colorful Manchurian windows had been made.

Chromatic glass etching is also a combination of Chinese and Western techniques. The artisans used Western techniques such as etching, grinding

夏至

Summer Solstice

June / 2021.6

21

星期一

Monday

辛丑牛年 甲午月

五月十二

or sandblasting to process the imported glass materials, while the content and drawing methods followed the traditional Chinese culture. Red, yellow, blue, green, purple, gold and other colors were used in glass etching. The frames and panes of different shapes, and the divided and matched colors would make the windows bright, transparent, luxurious and elegant. These colourful glass windows are the treasure of Lingnan architectural art and one of the important cultural symbols of Cantonese culture.

西关大屋的满洲窗。（图片来源：FOTOE 图片库）

The Manchurian windows of a grand house in Xiguan area. (The picture is bought from the photo gallery FOTOE.)

June / 2021.6

22

星期二

Tuesday

辛丑牛年 甲午月

五月十三

妙手夺天工

Excellent Craftsmanship

十三行时期，西方对于具有中国意象的各种精美器具和工艺品需求量很大，使得其在中国对外出口贸易中所占比重很大，由此促进了广东手工艺的发展。全盛时期，十三行一带有 5000 余家专营外销商品的店铺，约 25 万匠人专门从事外销工艺品的生产和制作，涉及漆器、银器、瓷器、纺织、绘画、雕刻等各个行业。

十三行的外贸传统拓宽了商人和工匠的眼界和见识。他们站在中国对外贸易的最前沿，感受着“西风”的吹拂，从商业与时代的需求中获取灵感，把欧洲绘画、磨花玻璃、珐琅彩、自鸣钟等工艺与本地传统结合并加以创新，形成了独具魅力的十三行工艺品。

During the Canton trade period, the Western countries had a large demand for all kinds of exquisite instruments and handicrafts with Chinese images, which took up a large proportion in China's export trade and promoted the development of the handicraft-making in Guangdong province. In the heyday of the trade, there were more than 5000 stores selling export commodities and about 250,000 craftsmen specializing in the production and processing of export handicrafts, such as lacquer wares, silver wares, porcelain, textile, paintings, sculptures, and products from other industries.

The tradition of foreign trade broadened the merchants' vision and insight, as well as those of the craftsmen. Standing at the forefront of China's foreign trade, feeling the tendency of Westernization, they sought inspiration from the demand of commerce and their epoch, so as to innovate the local tradition by integrating

June / 2021.6

23

星期三

Wednesday

辛丑牛年 甲午月

五月十四

the European techniques of painting, making polished glass, enamel wares, chiming clocks and other artifacts, and endowed the handicrafts sold in the trading region with a unique charm.

清代外销画：广州十三行漆器店。（图片来源：FOTOE 图片库）

A shop that sold lacquer wares in the trading region, Qing dynasty. (The picture is bought from the photo gallery FOTOE.)

June / 2021.6

24

星期四

Thursday

辛丑牛年 甲午月

五月十五

粤工入禁城

The Cantonese Craftsmen Entered into the Qing Court

十三行不仅向清皇室供应来自西洋的奇珍异宝，也为清宫源源不断地输送着顶尖的匠人。例如，清代宫廷十分青睐西洋玻璃制品，康熙年间，十三行的工匠程向贵、周俊二人，率先掌握了欧洲磨花玻璃技术，被召入内廷玻璃厂，制造出中外玻璃技术融合的作品“雨过天晴刻花套杯”，达到了当时工艺美术的最高水平。此外，供职内廷的不少珐琅技师也来自十三行；清宫造办处自鸣钟的技术骨干除了西洋人就是广东人。广州牙雕工匠的技术水平在雍正时已胜过苏州工匠，广州陈祖章父子等于 1741 年雕制出牙雕珍品《月曼清游册》。广东的木器工艺在工匠进入宫廷后声名大振，1736 年，造办处内正式成立了“广木作”，广东木作从此成为享誉全国的一个主流派别。

The Thirteen Hongs not only provided the Qing royal family with rare and ingenious objects from the West, but also sent top craftsmen to the court. For example, the Qing court was very fond of the glass wares from Europe, so during the reign of Emperor Kangxi, Cheng Xianggui and Zhou Jun, the craftsmen of the Thirteen Hongs, took the lead in mastering the European glass-polishing technology, and were called to work for the glass factory of the inner court, applying the Chinese and foreign technologies to produce a set of glass wares called “cups with carved flowers symbolizing the sunny weather after the rain”, which had reached the highest level of arts and crafts by that time. In addition, many enamel technicians working in the inner court also came from the Thirteen Hongs; the core technicians who made chiming clocks in Zaobanchu, the royal workshop of the Qing court, were either Cantonese or Westerners. The Cantonese ivory carvers' skill was better than that of the carvers from Suzhou in the period of Emperor Yongzheng. Chen Zuzhang

June / 2021.6

25

星期五

Friday

辛丑牛年 甲午月

五月十六

and his son in Canton produced the precious carved ivory work "Roaming under the Moon" in 1741. Guangdong's wood craft became famous after it had been introduced into the court. In 1736, the royal workshop officially established "Guangmuzuo" (workshop of Cantonese woodcraft). Since them, Guangdong's woodcraft had become a leading technique renowned in the country.

June / 2021.6

26

星期六

Saturday

辛丑牛年 甲午月

五月十七

清代檀香木雕庭院人物纹名片盒。檀香木质，盒身正反两面剔地浮雕亭台楼阁、人物，盖边和盒底边饰花卉、瓜果纹。图案精雕细刻，打开盒盖，仍有余香。

Sandalwood busuness card box with patterns of gardens and figures, Qing dynasty. The box is made of sandalwood, with the outer and inner sides carved with pavilions, and figures. The cover and bottom of the box are decorated with patterns of flowers, melons and fruits, which are exquisitely carved. When the cover is opened, there is still a fragrance.

清代贝壳彩绘罗汉图摆件。这对摆件采用完整的贝壳制成，以具有天然亮丽光泽的内壁为地，彩绘九名罗汉，并饰有各类树木、山石、花鸟。人物描绘细腻，富有动态，颜色鲜艳，配有随形定制的红木梅枝底座。

Carved shell with polychrome arhats, Qing dynasty. The pendants are made of two shells in whole. On the inner side of the shell that bears a natural bright luster, nine Arhats are painted and decorated with various trees, rocks, flowers and birds. The figures are delicate, colorful, with vivid gestures. Each shell is placed in a rosewood base customized in the shape of plum branches.

June / 2021.6

27

星期日

Sunday

辛丑牛年 甲午月

五月十八

潘壶出粤人

Teapots of Pan Style

19 世纪上半叶，一种名为“潘壶”的宜兴紫砂壶精品突然横空出世，风靡一时。潘壶壶身圆润，装饰极少，常于盖沿上以阳文篆字“潘”作为落款，整体呈现出大气、简练、明快的艺术风格，按形制分为“矮潘”（壶身扁柿形）、“中潘”（壶身扁球形）和“高潘”（近梨形）三种。

潘壶的设计定制者是十三行行商潘仕成，其不俗的文化品位和极高的艺术审美令潘壶身价不俗。据说在潘仕成老家福建莆田，还由此衍生出一个婚俗——在女儿出嫁时必以一潘壶为嫁妆，寓意期盼女子能够旺夫，保佑家族兴旺发达。潘壶如今已是紫砂壶的一个经典品类，但 200 年前的潘壶大多随潘家的衰败而流散，目前传世品不多。

In the first half of the 19th century, the “Pan's Teapots”, a kind of dark-red enameled pottery made in Yixing city were invented and suddenly became popular. The Pan's Teapot has a round shape with few decorations. It is often signed with a seal character “Pan” on the edge of the lid. It presents an overall artistic style of decency, simplicity and lightness. According to different shapes, the teapots can be divided into three types: “Short Pan” (which is flat like a persimmon), “Middle Pan” (which is like a flatten ball) and “High Pan” (which is like a pear).

The Pan Teapots were designed and customized by Pan Shicheng, a Hong merchant, whose excellent cultural taste and superior aesthetics made the teapots valuable. A marriage custom was related to the teapots of this kind in Putian, Fujian province, where Pan Shicheng was born. When someone's daughter get married, she would take a Pan's Teapot as her dowry, implying that she would bring luck to her

June / 2021.6

28

星期一

Monday

辛丑牛年 甲午月

五月十九

husband and prosperity to her family. Pan Teapots now have been a classical type of dark-red enameled pottery, but most of them were lost 200 years ago with the decline of the Pan family. At present, only few of them can still be found.

June / 2021.6

29

星期二

Tuesday

辛丑牛年 甲午月

五月二十

潘壶的设计定制者十三行行商潘仕成，祖籍福建，世居广州，是晚清享誉朝野的官商巨富。（图片来源：FOTOE 图片库）

The designer of "Pan's Teapots", Pan Shicheng, whose ancestors were from Fujian province, was a wealthy official merchant in Canton, renowned both in the court and in society. (The picture is bought from the photo gallery FOTOE.)

新奇外销画

Export Paintings Of New Style

18世纪中期到20世纪初，为了满足西方消费者对“中国趣味”风尚的追捧，在广州十三行商区出现了一批中国画师。他们将中国传统绘画技法与西洋技法，如透视法、色彩晕染、绘画形式和颜料画材等相结合，以富有中国情调的人像、风俗、庭院、花鸟、植物、自然风光、城市景观、船只港口、生产交易等为题材，创制出一种既不同于传统中国画，又与道地西洋画有所差异的美术作品，主要行销海外，因此被称为“外销画”。

外销画是一个相对宽泛的概念，包括油画、玻璃画、水彩画、水粉画、通草画等，由于作品极富“中国趣味”，深受西方人士欢迎。据说，在外销画最繁荣的时候，仅十三行地区就有30家画馆、两三千名画师。

From the middle of the 18th century to the beginning of the 20th century, in order to satisfy the Western consumers' pursuit of "Chinoisere", a group of Chinese painters appeared around the trading region. They combined the traditional Chinese painting techniques with the Western ones, such as perspective, sfumato, pictorial forms, pigments and other materials, and created a kind of artworks different from traditional Chinese paintings or native Western paintings with the themes of portraits of the Chinese people, customs, courtyards, flowers and birds, plants, natural scenery, urban landscapes, ships, ports, and scenes of production and trade, etc. These art works were mainly sold overseas, so they were called "export paintings".

Export paintings refer to a relatively broad concept, including oil paintings, glass paintings, watercolor paintings, gouache paintings, pith paintings, etc. Because the works were full of "Chinoisere", they were popular among the Westerners. It is said that at the prime of the export paintings, there were 30 galleries and 2,000 to 3,000 painters working in trading region.

June / 2021.6

30

星期三

Wednesday

辛丑牛年 甲午月

五月廿一

July

一	二	三	四	五	六	日
			1 建党节	2 廿三	3 廿四	4 廿五
5 廿六	6 廿七	7 小暑	8 廿九	9 三十	10 六月	11 初二
12 初三	13 初四	14 初五	15 初六	16 初七	17 初八	18 初九
19 初十	20 十一	21 十二	22 大暑	23 十四	24 十五	25 十六
26 十七	27 十八	28 十九	29 二十	30 廿一	31 廿二	

绘于1855年的纸本水粉画：关联昌（庭呱）画室。关联昌是当时最多产的外销画画师之一。（图片采自《东西汇流》）

The studio of Tinqua, gouache painting on paper, 1855. Tinqua, Guan Lianchang was the most prolific export painters at that time. (The picture is taken from *East Meets West*.)

July / 2021.7

1

星期四

Thursday

辛丑牛年 甲午月

五月廿二

建党节

The Party's Day

飞入百姓家

Export Paintings in Ordinary Families

外销画在民间商业需求中产生，自由、自发、草根，有着强烈的生命力和传播力。

十三行外销画历史上有几个相当有名的外销画师，他们学习西洋绘画技法的途径主要有出洋学习、当来华外国画家的学徒和外销画师之间互相传授三种。但更多的无名画师仅仅通过模仿或得到零散的指点便开始从事绘画工作，他们中的很多人只掌握了最简单的一点儿皮毛知识，因而所绘的外销画品质并不高，被当成廉价的旅游纪念品或贸易品，以量大取胜。英国美术史家苏立文曾如是评价外销画："像沙子那样掉入下层专职画家和工匠画家的手里。"点出了当时外销画水平参差不齐的状况。

The export paintings were made to cater to people's commercial demand, so they were marked by the spirit of freedom, spontaneity and humbleness, full of strong vitality and easily disseminated.

There are several well-known export painters in the history of export paintings of the Canton trade. There were mainly three ways for them to learn the Western painting techniques: to study abroad, to be apprentices of foreign painters in China and to learn from each other. But there were more unknown painters who started to paint by imitating or getting sporadic instructions. Many of them had only learned the simplest and plainest knowledge about painting, so the quality of the exported paintings painted by them were not high, and regarded as cheap souvenirs or trade goods only to be sold in large quantities for profits. Michael Sullivan, a British art historian, once commented on the export paintings as follows: like sand that drifts, they fell into the hands of the specialized painters and craftsmen of the lower level. This points out the uneven qualities of the export paintings at that time.

July / 2021.7

2

星期五

Friday

辛丑牛年 甲午月

五月廿三

July / 2021.7

3

星期六

Saturday

辛丑牛年 甲午月

五月廿四

清末泥塑彩绘女像。

Clay sculpture of a woman, late Qing dynasty.

July / 2021.7

4

星期日

Sunday

辛丑牛年 甲午月

五月廿五

清末泥塑彩绘男像。

Clay sculpture of a man, late Qing dynasty.

追随钱纳利

Following George Chinnery

英国画家乔治·钱纳利虽不是严格意义上的外销画画家，但他的画风对中国外销油画的发展影响巨大。19 世纪中期，广州著名的外销画家林呱（关乔昌）就是钱纳利的弟子。钱纳利以“华丽风格”著称，尤其擅长女性肖像画，在表现上以奔放的笔触、明亮的色彩和强烈的对比为特征，这些特征在现存很多十三行外销画中仍能找到。

钱纳利于 1825 年到达澳门，此后在中国华南地区居住了 20 余年，直到去世。钱纳利在澳门开设了画室，为欧洲人和中国显贵绘制画像，时常往来于澳门、广州、香港三地，留下了大量反映市井街头的速写、素描、油画和水彩作品。受其影响，中国画师绘制的外销油画也多以肖像画和港口风景为主。

Although the English painter George Chinnery was not a painter of export paintings in a strict sense, his painting style had a great influence on the development of Chinese export paintings. In the mid-19th century, Lamqua (Guan Qiaochang), a famous export painter in Canton, was a disciple of George Chinnery who was famous for his “gorgeous style”, and especially good at female portraits. His expression was characterized by bold strokes, bright colours and strong contrast, which can still be found in many export paintings of the Canton trade period.

George Chinnery arrived at Macao in 1825 and lived in South China for more than 20 years until his death. He opened a studio in Macao to paint portraits for European and Chinese dignitaries. He often traveled to Macao, Canton and Hong Kong, leaving behind a large number of sketches, oil paintings and watercolour works whose themes were streets and daily life of the local resident. Influenced by him, Chinese painters also took portraits and port landscapes as their main subjects.

July / 2021.7

5

星期一

Monday

辛丑牛年 甲午月

五月廿六

乔治·钱纳利晚年自画像。钱纳利的画作，客观上对中国油画的发展起到促进作用，他将19世纪的英国画风带入中国，在华南一带——特别是广州，广为传播。

A self-portrait of George Chinnery painted in his later years. The paintings of Geroge Chinnery played an extremely important role in the development of Chinese oil paintings. He introduced the British painting style of the 19th century into China and the style was widely spread in South China, especially in Canton.

July / 2021.7

6

星期二

Tuesday

辛丑牛年 甲午月

五月廿七

开山史贝霖

Spoilum, the Forerunner

史贝霖（根据画作上的署名“Spoilum”音译，真实中文名字不可考），是目前已知最早的广州外销画家，大约活跃于清代乾隆、嘉庆年间，以擅长绘制玻璃油画肖像而出名。

史贝霖有着娴熟高超的西洋绘画技法，从 1786 年起，他的创作自玻璃油画转向布面油画，渐渐形成强烈的个人风格，比如人物背光面的背景有浅色光晕，或者人物的脸部从侧上方受到光影投射，使得脸部与背景拉开了距离。这种风格被称为“史贝霖画风”，影响了一大批外销画师。史贝霖的代表作有早期玻璃油画《托马斯·弗雷船长肖像》（1774 年）及晚期布面油画《行商肖像》（1800 年）等。

Spoilum (this is only a transliteration of the signature “Spoilum” on the painting, and his real Chinese name is untraceable), was the earliest known export painter in Canton, active in the period of Emperor Qianlong and Jiaqing in the Qing dynasty, and famous for his expertise in oil paintings on glass.

Spoilum had excellent Western painting skills. Since 1786, his focus changed from oil painting on glass to oil painting on canvas, gradually forming a strong personal style. For example, he added a light glow around the figure on the background, or projected the light on the figure’s face from one side or above, forming a distance between the face and the background. This style, known as the “Spoilum’s style”, had influenced a large number of export painters. His representative works include an oil painting on glass “Portrait of Captain Thomas Frey” in his early period (1774), and an oil painting on canvas “Portrait of a Hong Merchant” in his later period (1800).

小暑

Slight Heat

July / 2021.7

7

星期三

Wednesday

辛丑牛年 乙未月

五月廿八

广州外销画画师史贝霖笔下的广州丝绸商人。（左图）

A Cantonese silk merchant, painted by Spoilum. (left)

史贝霖画作，油画《广州商馆区》。（右图）（图片来源：FOTOE 图片库）

The Trading Region in Canton, an export painting by Spoilum. (right) (The picture is bought from the photo gallery FOTOE.)

July / 2021.7

8

星期四

Thursday

辛丑牛年 乙未月

五月廿九

玻璃画：舶来又舶去

Reverse Glass Paintings: the Ships are Coming and Going

玻璃画最早出现于 15 世纪的意大利，大约在 17 世纪与其他玻璃制品一起作为外交礼物被带入中国。玻璃画在中国兴盛于 18 世纪乾隆和嘉庆时代，由于在镜子背面作画，且描绘方向与一般画法相反，因此这门技术也叫作“背画技术”。

绘制玻璃画的工匠首先要按草图把镜子背面的水银镀膜铲掉，然后用油彩、水粉或金粉依照从里到外的步骤上色画画，把被铲掉水银镀膜的部分填补完好。还有一种是直接在透明玻璃板上作画，画完后在背面封涂水银层固色保存，玻璃肖像画大部分属于这一类。在中国外销画师的创新下，传自海外的玻璃画因具有浓厚的东方风情而返销海外，成为 18 世纪外销画中最耀眼的品种，包括“玻璃镜镶玻璃油画”“银镶边玻璃油画盒”“镶油画玻璃盒”和“玻璃影画时辰钟”等类别。

Reverse glass paintings first appeared in Italy in the 15th century, and were brought into China as diplomatic gifts together with other glass products in the 17th century. Reverse glass paintings flourished in China during the reign of Emperors Qianlong and Jiaqing, namely the 18th century. Unlike other general method, the patterns are painted reversely on the back of the mirror, thus the paintings are called the “technique of painting on the back”.

To make the reverse glass paintings, the craftsmen first shoveled the mercury coating on the back of the mirror according to the sketch, and then painted with oil, gouache pigments as well as gold powder; they must paint from the inside to outside, and again fill in the parts where the mercury coating had been removed.

July / 2021.7

9

星期五

Friday

辛丑牛年 乙未月

五月三十

Another method was to paint directly on a transparent glass plate. After the painting had been done, it would be covered with a layer of mercury on the back for fixing the color and preservation. Most of the portraits on glass belong to this category. With the innovation of Chinese export painters, the reverse glass paintings, whose technique originated from overseas, were made with strong oriental aura to be sold abroad and became the most eye-catching one among the export paintings of the 18th century. These types include the "oil paintings on glass framed in mirror", "glass boxes with oil painting rimmed with silver", "oil paintings inlaid in glass box" and "clocks with glass painting".

July / 2021.7

10

星期六

Saturday

辛丑牛年 乙未月

六月初一

清末仕女肖像图玻璃画。画家采用反笔彩绘技法，在玻璃背面施以油彩绘制。画中拈花仕女低头自怜，眉清目秀。画匠对人物脸部、手部进行了细致的描绘，连发根都清晰可见，衣服纹饰精致明了，衣褶流畅，通过衣褶上的明暗色调变化的处理来表现其动态感和立体感。

Portrait of a maiden, reverse glass painting, late Qing dynasty. The painter painted reversely and applied oil pigments on the back of the glass. In the painting, the lady, who pinched a flower in her hair, tilted her head and seemed to pity herself. The painter has made a detailed depiction of the face and hands. Even the hairs are clearly visible. The creases of her clothes are delicate and clear, and the pleats are depicted smoothly. The dynamic and three-dimensional effect is expressed through the changes of light and shadow on the pleats.

July / 2021.7

11

星期日

Sunday

辛丑牛年 乙未月

六月初二

清末花鸟图玻璃画。一只开屏孔雀精神抖擞地立于湖石之上，四周禽鸟或闲庭信步觅食，或于枝头引颈高歌，一派春色满园的绚丽风光，四周用如意花卉纹作为装饰框。这幅玻璃画保存完好，画工精细，构图丰富饱满。

Flowers and birds, reverse glass painting, late Qing dynasty. A peacock in high spirits stands on the rock of the lake, surrounded by birds that are strolling around for food, or singing in the branches. It is a wonderful scenery of spring in full swing. The frame is decorated with Ruyi and floral patterns. This reverse glass painting is well preserved, with fine craftsmanship and a full composition.

中西两辉映

A Brilliant Contrast Between Chinese and Western Style

目前已知在广州创作的最早一幅玻璃肖像画是一位中国佚名画师绘制的《约翰·派克肖像》，作画时间不晚于 1744 年。画中的人物戴着披肩假发，穿着华丽的长外套和马甲，是当时欧洲肖像画的流行样式，而背景是极富中国文化特征的山水庭园。这种中西结合的造型样式几乎成为 18 世纪中国玻璃肖像画的通用形制：肖像主人公身穿奢华的欧洲服饰，全身或半身居于画面前景的偏左或偏右位置，置身于以中国山水、建筑、庭园、花草、植物等为背景的场景中。这种将人物写实与带有强烈装饰性的中国山水风景相结合的手法，迎合了当时西方商人、官员、水手的审美情趣，甚至引领了一波来华外国人绘制肖像画的风气。

The earliest portrait on glass known at present is *The Portrait of John Parker*, created in Canton by an unknown Chinese painter, and the time of production was not later than 1744. The figures in the painting were wearing wigs, gorgeous long coats and waistcoats, which was the popular style of European portraits at that time, while the background was a landscape of garden with Chinese cultural characteristics. This combination of Chinese and Western style almost became the general form of Chinese portraits painted on glass in the 18th century: the figure of the portrait is wearing luxurious European clothes, and his whole body or upper body is on the left or right side of the foreground in the picture, whereas in the background, there are Chinese landscape, architectures, garden, flowers and plants. The combination of the realistic figures and highly decorative Chinese landscape catered to the aesthetic taste of the Western traders, officials and sailors at that time, and even led to a fashion for foreigners to have their portraits painted in China.

July / 2021.7

12

星期一

Monday

辛丑牛年 乙未月

六月初三

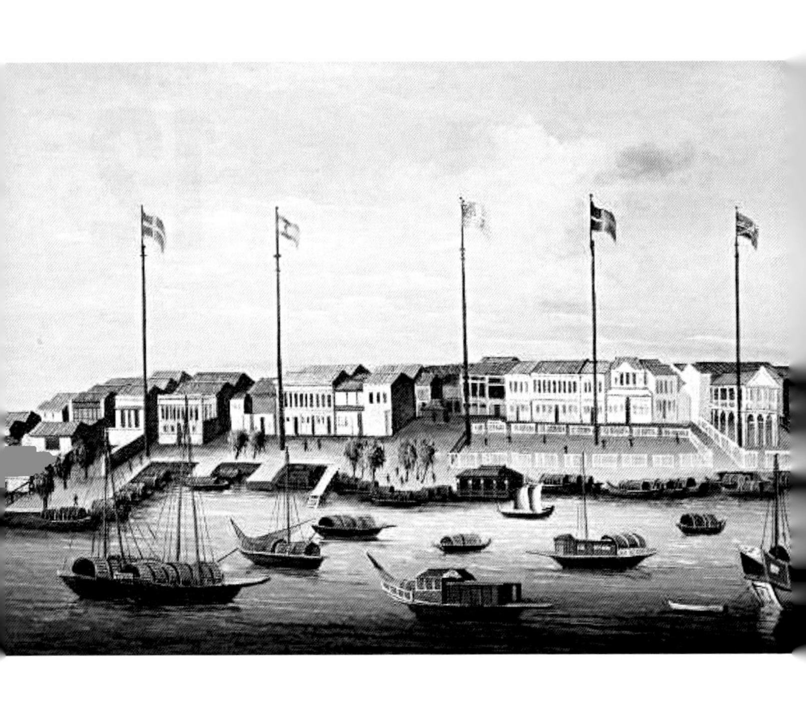

18世纪末至19世纪初的玻璃画：广州十三行商馆区。

A reverse glass painting of the late 18th or early 19th century. It depicts the trading region in Canton.

July / 2021.7

13

星期二

Tuesday

辛丑牛年 乙未月

六月初四

精明关乔昌

Lamqua, the Shrewd

关乔昌是 19 世纪中期广州最重要的外销画师。他早期从本地画师那里学到写实油画技巧，后跟随乔治·钱纳利学画，掌握了英国学院派肖像画技巧，在广州、澳门开设了画室，富有商业头脑的他在门牌广告上标明：“林呱，英国和中国的画家”，还在名片上自称“中国的托马斯·劳伦斯”，以招徕顾客。

关乔昌一生极为高产，其中既有粗劣的流水线产品，也不乏艺术水准极高的精品。1835 年，由其创作的《老人头像》入选英国皇家美术学院展，他也因此成为最早在欧洲画展上亮相的中国画家；流传至今的伍秉鉴画像也出自关乔昌之手；他还以同样的角度、姿势和手法为 30 岁和 52 岁的自己画了两幅自画像，浓重的黑色背景下，人物被一侧打来的光照亮，形成立体生动的阴影，展现了其对古典主义伦勃朗光线的纯熟运用。

Lamqua (Guan Qiaochang) was the most important export painter in Canton in the mid-19th century. In his early days, he learned the skill of realistic oil painting from the local painters, and later followed George Chinnery. He had mastered the British academic portrait style, and opened studios in Canton and Macao. With his commercial acumen, he wrote on the billboard on his door: “Lamqua, a British and Chinese painter”, and addressed himself as “Thomas Lawrence of China” on his business card to attract customers.

Lamqua was extremely prolific throughout his life. His works not only included products bearing the roughness of those made in assembly lines, but also high-quality fine pieces. In 1835, *Portrait of an Old Man*, painted by Lamqua, was selected into the exhibition held by Royal College of Art. He became the first Chinese painter

July / 2021.7

14

星期三

Wednesday

辛丑牛年 乙未月

六月初五

whose works appeared in European art exhibition. The popular portrait of Wu Bingjian was also painted by Lamqua. He also drew two self-portraits at the age of 32 and 52, respectively applying the same angle, posture and technique. Under the background of dense darkness, the figures are illuminated by the light from one side, forming realistic and vivid shadows. This shows that Lamqua was good at using the classical Rembrandt light.

July / 2021.7

15

星期四

Thursday

辛丑牛年 乙未月

六月初六

关乔昌自画像。（图片采自《清代洋画与广州口岸》）

The self portrait of Lamqua. (The picture is taken from *Western Paintings and the Canton Port during the Qing Period*.)

庭呱关联昌

Tinqua, Guan Lianchang

关乔昌的四弟关联昌也是有名的外销画师，英文名音译为庭呱，擅长的领域是水彩画和细密画，以成套的人物风俗画和生产风俗画著称。庭呱很懂得外销画经营之道。当时不少西方人很想了解中国茶叶、丝绸和瓷器等的生产过程以及中国风俗，庭呱就趁势推出了成套的风俗画，如描绘广州三百六十行的市井风情画，每套 120 幅，内容包括锁鞋边、执字纸、算命、钉屐、卖葵扇等。

由于工作量太大，庭呱在十三行同文街 16 号开了一间画室，雇佣学徒和画师像流水线工人般从事创作。绘制于 1855 年的水粉画《庭呱画室》以从外向里看的视角，记录下了当年庭呱工作室的实景和画工的工作状态。庭呱存世画作很多，其代表作为《插秧》，收藏于美国迪美美术馆。

Lamqua's fourth younger brother, Guan Lianchang, was also a famous export painter, known as Tinqua. He was good at watercolor painting and miniature, and famous for making painting albums of customs, figures and production scenes. Tinqua knew how to sell export paintings. At that time, many Westerners were eager to understand the production process of Chinese tea, silk and porcelain as well as Chinese custom, and Tinqua took advantage of the trend to launch complete sets of custom paintings, such as those depicting all kind of professions in Canton. Each set consists 120 pieces, depicting people mending shoes, collecting paper with written characters, telling fortune, nailing clogs, and selling fans made from sabal leaf, etc.

Due to the heavy workload, Tinqua opened a studio at No. 16, Tongwen Street in the trading region and hired apprentices and painters to work in assembly line. *Tinqua's Studio*, a gouache painting created in 1855, depicts the actual scene of Tinqua's studio and the working situation of the painters from a perspective

July / 2021.7

16

星期五

Friday

辛丑牛年 乙未月

六月初七

of looking from the outside. Many paintings of Tinqua can still be found. A representative work is called *Rice Transplanting*, collected in Peabody Essex Museum in the United States.

19世纪末清代外销通草画：商贩图。

Workers and vendors, watercolour on pith paper, late 19th century.

July / 2021.7

17

星期六

Saturday

辛丑牛年 乙未月

六月初八

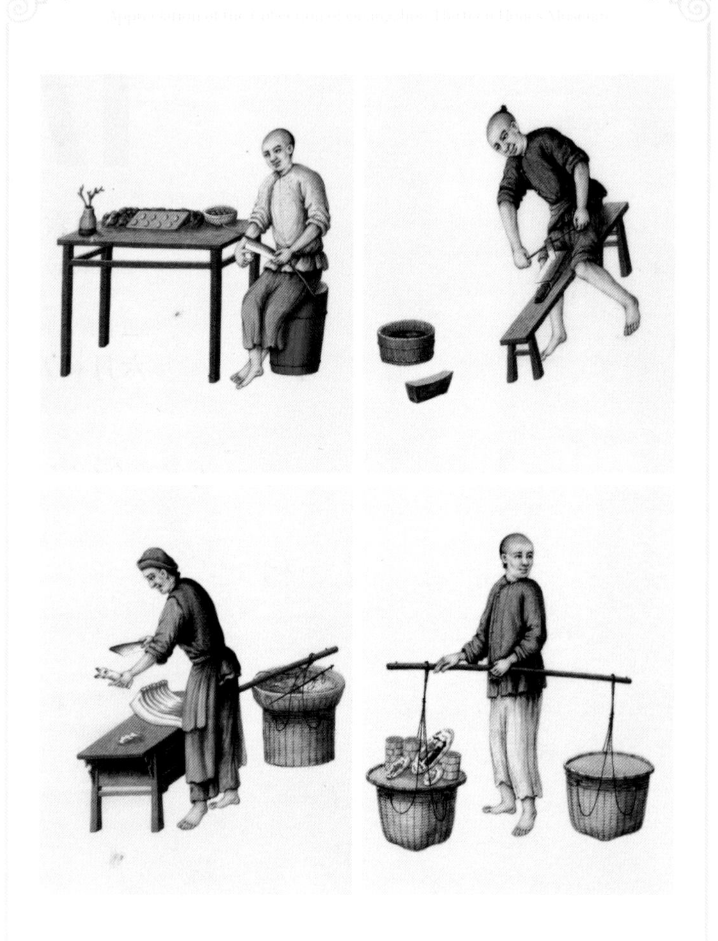

19世纪末外销通草画：商贩图。

Workers and vendors, watercolour on pith paper, late 19th century.

July / 2021.7

18

星期日

Sunday

辛丑牛年 乙未月

六月初九

象牙画细密

Ivory Miniatures

细密画是一种小型绘画，始于《古兰经》的边饰图案，是波斯艺术的重要门类，以精细刻画著称，常常绘于书籍、徽章、匣子、象牙板、铜板或黄铜板上，作为肖像画或装饰画。细密画是十三行外销画中相对高端的品种，结合了中国传统山水画技法，主要画于瓷器、灯彩、书签等上面，不但有绘画，而且有刻、镶等多种形式，与波斯细密画有所区别。在十三行外销画师中，关联昌极为擅长绘制细密画，他绘制的玛尼哥特肖像曾被美国学者卡尔·克罗斯曼作为经典案例录入其著作《中国外销装饰艺术：绘画、家具与珍玩》。

The miniatures are a kind of small paintings, which originated from the decorative border pattern of *the Koran*. As an important category of Persian Art and famous for their fine depiction, they are often painted on books, coat of arms, caskets, ivory plates, copper plates and brass plates as portraits or decorative paintings. The miniatures, as a relatively high-end category of export paintings in Canton trade, were integrated into the traditional Chinese landscape painting techniques. They were mainly painted on porcelain, lamps, bookmarks, etc. There were not only paintings, but also various forms of engraving and inlaying, which were different from the Persian miniatures. Among the export painters in the trading region, Tinqua was especially good at drawing miniatures. The "Portrait of Louis Manigault" painted by him was shown as a typical example in the book *The Aecorative Arts of the China Trade: paintings, furnishings and exotic curiosities* by an American scholar Carl L. Crossman.

July / 2021.7

19

星期一

Monday

辛丑牛年 乙未月

六月初十

清代外销水粉画：广州十三行贸易码头。关联昌绘于 1855 年。（图片来源：FOTOE 图片库）

A gouache painting of the Qing dynasty: the wharf in the trading region in Canton, painted by Tinqua, 1855. (The picture is bought from the photo gallery FOTOE.)

July / 2021.7

20

星期二

Tuesday

辛丑牛年 乙未月

六月十一

挂纸成意趣

The Hanging Paper Paintings

18 世纪早期，广州口岸的外销画主要为纸本绘画和彩色木版画。为了方便运输，并尽可能多增加装载数量，纸本绘画通常未经装裱，只在背面衬一张背纸，也没有画框或轴头，可以直接悬挂或张贴在墙上，故得名“悬挂纸画”。

悬挂纸画的内容大多是中国的风土人情，也有表现富裕家庭的快乐闲适生活，以及农民和手工业者的生产生活。它们通常成套制成，形成一系列连续的景象，可以贴满整个房间。17 至 18 世纪，这种既富有“中国趣味”，又价格低廉的悬挂纸画在欧洲十分流行，是当时相当时髦的房间装饰品，有人直接拿来当壁纸使用，因此也叫“壁纸纸画”。

In the early 18th century, the paintings exported from Canton were mainly painted on paper or made by xylography. In order to facilitate the transportation and increase the loading quantity as much as possible, the paper paintings, usually not mounted, would be backed by only a piece of paper, without frames or spindles, and could be directly hung and posted on the wall, so they were called “hanging paper painting”.

The themes of hanging paper paintings are mostly the local customs of China, showing the happy and leisure life of the rich families, as well as the working life of farmers and craftsmen. They are usually made in sets, forming a series of continuous scenes that can be pasted all over the room. In the 17th and 18th century, the cheap hanging paper paintings with “Chinese taste” were very popular in Europe. It was a very fashionable room decoration at that time. Some people directly used it as wallpapers, so they were also called “wallpaper paintings”.

July / 2021.7

21

星期三

Wednesday

辛丑牛年 乙未月

六月十二

19世纪清代外销通草画：水彩人物图。（原件藏于广州十三行博物馆）

Figures, watercolour on pith paper, export painting, 19th century. (The original piece is collected in Guangzhou Thirteen Hongs Museum.)

大暑

Great Heat

July / 2021.7

22

星期四

Thursday

辛丑牛年　乙未月

六月十三

通草失复得

The Rediscovery of Pith Paintings

通草画是外销水彩画中的一种，与普通水彩画的区别在于用纸的不同。通草纸是用一种学名为通脱木（通草）的灌木茎髓切割而成，每张通草纸不过人的两三只巴掌大小，但由于其质地经水彩着色后，能够在光折射下呈现色彩缤纷的效果，可媲美漆器或刺绣，故深受画师和消费者喜爱。通草画题材以反映清末广州的社会生活场景和各种形色人物为主，被称为“广州明信片”。

通草画出现于1820年，19世纪30至60年代为鼎盛期，画工达两三千人。但随着摄影术和制版印刷术的出现，外销画渐趋衰落，通草纸的制作方法也一度失传。直到21世纪初，广州市博物馆在经过大量的搜集、研究和实地走访后，从贵州省贵定县引回通草纸制作技术，令通草画重现于世。

The pith paintings are one of the watercolor paintings for export. The pith painting's paper is different from that of the traditional Chinese paintings. Pith paper is made of the pith of a shrub known as Tongtuomu (Tongcao). Each piece of pith paper is cut into small size of no more than two or three palms. When it is painted by watercolour pigments and placed under the refraction of light, its texture will take on a colorful effect, which is comparable to that of the lacquer wares or embroidery, so it is popular with painters and consumers. The themes of the pith paintings mainly reflect the social life and various characters in the late Qing dynasty, and they are known as “the postcard of Canton”.

The pith paintings appeared in 1820s, and prevailed from 1830s to 1860s, during which there were two or three thousand painters. However, with the emergence of photography and plate printing, the export paintings gradually declined, and the production method of pith paper was lost for a time. It was at the beginning of the

July / 2021.7

23

星期五

Friday

辛丑牛年 乙未月

六月十四

21st century that Guangzhou Museum re-introduced the technology of producing pith paper from Guiding county, Guizhou province after a large amount of research, study and fieldwork, which brought to a rediscovery of the pith paintings.

19世纪清代外销通草画：水彩昆虫图。

Insects, watercolour on pith paper, export painting, 19th century.

July / 2021.7

24

星期六

Saturday

辛丑牛年 乙未月

六月十五

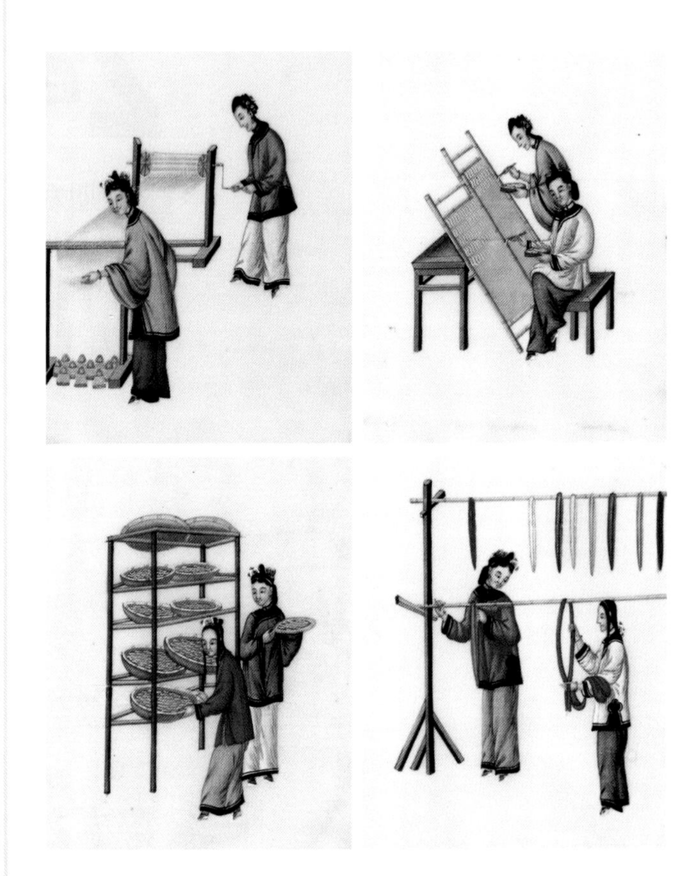

19世纪外销通草画：采桑、养蚕、缫丝、织布图（局部）。

The selected scenes the process of making silk,watercolour on pith paper, 19th century.

July / 2021.7

25

星期日

Sunday

辛丑牛年 乙未月

六月十六

本土博物画

The Local Paintings of Natural Subjects

植物类绘画是早期外销画常见题材，目前已知最早的一幅植物外销画绘制于 1700 年。与传统中国画写意动植物相比，外销画中的动植物绘画讲求描绘精准，近似于西方博物画，甚至于在绘制花卉整体时，还会附上花朵结构分解图。事实上，动植物外销画确实受到 18 世纪西方博物学兴起的影响。当时不少外国传教士进入中国大量采集和搜集动植物标本和信息，为了进行精确记录，对动植物绘画的科学和精准有着很高要求，因而推动了动植物外销画的发展。不过随着后期外销画商业模式的成熟，流水线的生产方式让动植物画渐渐偏离了写实性与规范性，不再具有科学研究意义。

The plants are a common theme of early export paintings. At present, the earliest known export painting of plants was painted in 1700. Compared with those of the traditional Chinese freehand brushwork paintings, the animals and plants in the export paintings are depicted with accuracy, similar to the paintings of natural themes in the West. A diagram showing the composition of a flower might even be attached to a painting which shows a flower as a whole. In fact, the export paintings of animals and plants were influenced by the rise of Western natural science in the 18th century. At that time, many foreign missionaries came to China to conduct research and collect a large number of animal and plant specimens and information. In order to record them accurately, they had high requirements for the paintings of animals and plants in terms of science and accuracy, which pushed forward the development of export painting of these themes. However, with the maturity of the business model of export paintings, the assembly line production gradually made the paintings deviated from realism and standardization, so the paintings no longer bore significance of scientific research.

July / 2021.7

26

星期一

Monday

辛丑牛年 乙未月

六月十七

19世纪清代外销通草画：水彩花蝶草虫图。（原件藏于广州十三行博物馆）

Flowers and butterflies, watercolour on pith paper, 19th century. (The original piece is collected in Guangzhou Thirteen Hongs Museum.)

19世纪清代外销通草画：水彩鱼图。（原件藏于广州十三行博物馆）

Fish, watercolour on pith paper, 19th century. (The original piece is collected in Guangzhou Thirteen Hongs Museum.)

July / 2021.7

27

星期二

Tuesday

辛丑牛年 乙未月

六月十八

包装月份牌 Ⅰ

The Calendar Posters Ⅰ

19世纪40年代，随着摄影术的传入，更加逼真、快捷、廉价的照片逐渐取代了外销画，广州外销画的质量和产量明显下降。为了适应新形势和新需求，很多优秀的外销画师将画室从广州搬到新崛起的上海、香港等地，但仍无法扭转行业颓势。部分外销画师开始兼营摄影，还有一部分则采用新的石印技术，开始制作19世纪末日益流行的仕女月份牌画和商标画。这部分画师中最有名的是关作霖的曾孙关蕙农，他是最早绘制月份牌画的画家，有“月份牌画王”之称。关蕙农幼时学习西洋画，后师从居廉，熔中西画法于一炉。他于1908年首创以西画油彩绘中国仕女，对后来美女月份牌绘画影响很大。

In 1840s, the introduction of photography, which was more realistic, convenient and cheaper, gradually replaced the export paintings, and the quality and output of Cantonese export paintings decreased significantly. In order to adapt to the new situation and demand, many skillful export painters moved their studios from Canton to the newly emerging market places such as Shanghai and Hong Kong, but still could not recover from the decline of the industry. Some of the export painters began to take photos, and at the end of the 19th century, some of them even began to use a new technology, lithography, to produce the popular calendar posters and trademark paintings of ladies. The most famous painter was Guan Huinong, Guan Zuolin's great grandson. He was the first painter to produce calendar posters, and was known as "the king of calendar posters". When Guan Huinong was a child, he studied Western painting. Later, he learned from Ju Lian (a famous painter in Guangdong province) and integrated the Chinese and Western painting techniques. In 1908, he first painted the Chinese ladies with Western oil pigments, which had a great influence on the calendar posters of beauties in later period.

July / 2021.7

28

星期三

Wednesday

辛丑牛年 乙未月

六月十九

关蕙农绘于1933年的月份牌。（图片来源：FOTOE图片库）

A calendar poster of 1933, painted by Guan Huinong. (The picture is bought from the photo gallery FOTOE.)

July / 2021.7

29

星期四

Thursday

辛丑牛年 乙未月

六月二十

包装月份牌 Ⅱ

The Calendar Posters Ⅱ

月份牌画是指将商品宣传广告与中国年画中配有月历和节气的“历画”结合起来形成的一种新年画，由商家出资向画家订购印刷，随商品附赠顾客，起到广告宣传、提高品牌知名度的作用。

月份牌画兴起于外销画衰落之后，主要流行于上海和香港地区，其题材有仕女、娃娃、历史故事、戏曲人物、名胜古迹五类，但这些内容只是用来吸引观众的注意力，一张月份牌画的“画眼”是处于画面一隅的商品图像和广告文字，这才是商人希望观众获取的有效信息。月份牌画宣传的商品丰富多样，包括香烟、火油、肥皂、蚊香、布料、化妆品、酒类、药品乃至肥田粉等。

The calendar posters were a kind of new calendar paintings which combined the commercial advertisements with the Chinese New Year's paintings, namely "calendar paintings with solar terms of each month". Printed by the painters, the paintings were ordered by the merchants to be attached to the commodities bought by the customers, playing the role of advertising and raising people's attention to the brands.

The calendar posters became popular in Shanghai and Hong Kong after the decline of export paintings. Their themes include Chinese ladies, dolls, historical stories, opera figures, and famous historical and cultural sites. But these contents were only used to attract the consumers' attention. The "most significant part of a calendar poster" was the commodity together with the advertising words in the corner of the picture, the efficient information to which the merchants wanted the customers to obtain. Various kinds of commodities were advertised in the calendar posters such as cigarettes, kerosene, soap, mosquito-repellent incense, cloth, cosmetics, alcohol, drugs, fertilizer, etc.

July / 2021.7

30

星期五

Friday

辛丑牛年 乙未月

六月廿一

19世纪清代外销通草人物画。

Figures, watercolour on pith paper, 19th century.

July / 2021.7

31

星期六

Saturday

辛丑牛年 乙未月

六月廿二

August

一	二	三	四	五	六	日
						1 建军节
2 廿四	3 廿五	4 廿六	5 廿七	6 廿八	7 立秋	8 七月
9 初二	10 初三	11 初四	12 初五	13 初六	14 七夕节	15 初八
16 初九	17 初十	18 十一	19 十二	20 十三	21 十四	22 中元节
23 处暑	24 十七	25 十八	26 十九	27 二十	28 廿一	29 廿二
30 廿三	31 廿四					

19世纪末通草水彩船舶图。

Ships, watercolour on pith paper, 19th century.

19世纪末通草水彩“两广部堂”船舶图。

A ship with a flag that says “the government of Guangdong and Guangxi provinces”, watercolour on pith paper, 19th century.

August / 2021.8

1

星期日

Sunday

辛丑牛年 乙未月

六月廿三

建军节

Army Day

广州初入镜

The Earliest Photos of Canton

1839年，摄影术诞生，几年后随着外国摄影术传入中国，迅速取代了外销画的地位。1844年，法国人于勒·埃及尔作为法国外交使团的一员，带着达盖尔相机到达澳门，第一次把摄影术传入中国，并在广东拍摄了40余张最早记录真实中国的照片。于勒在参与起草中法《黄埔条约》期间，为耆英拍了一张人物肖像，耆英也因此成为第一个被拍“小照”的中国人。于勒的另一张照片以广州全景为主题，记录了1844年广州的城市风貌。

皮埃尔·约瑟夫·罗西耶则是目前已知最早在欧洲发行中国题材商业照片的摄影师。他在广州拍摄了一批照片，内容涵盖中外人像和城市风景，如广州南城门、官员内宅、五百罗汉堂、贡院庭院、广州花塔、财政街等，其中很多建筑早已消失，更显得这批存世照片的珍贵。

Photography, born in 1839, was introduced into China by foreign photographers and quickly took the place of export paintings. In 1844, as a member of the French diplomatic corps, Jules Itier arrived at Macao with a daguerreotype camera. He was the first photographer who had introduced photography into China. The 40 photos of Guangdong taken by him were the earliest photos taken in China. During his participation in the drafting of the *Treaty of Whampoa* between China and France, Jules Itier took a portrait of Qi Ying, who became the first Chinese to be photographed. Another photo taken by him shows a panorama of Canton, and records the city's scenery in 1844.

Pierre Joseph Rossier was the first photographer known to release photo albums of Chinese commercial themes in Europe. He took a number of photos in Canton, including portraits of local and foreign people and urban landscapes, such

August / 2021.8

2

星期一

Monday

辛丑牛年 乙未月

六月廿四

as the Southern Gate of Canton, the officials' houses, Hall of Five Hundred Arhats, Gongyuan Courtyard, Flower Tower of Canton, Financial Street, etc. Many of the architectures in the photos have long disappeared, so the still existing photos are especially precious.

于勒·埃及尔拍摄的广州，摄于 1844 年。

A photo of Canton by Jules Itier, 1844.

August / 2021.8

3

星期二

Tuesday

辛丑牛年 乙未月

六月廿五

外销瓷：中华绝密技 I

Export Porcelain: China’s Secret Technology I

中国陶瓷作为外销产品的历史很悠久。据考古发掘证明，至迟到唐代，陶瓷已经作为外销品出现在“海上丝绸之路”的商船里。大致而言，广东地区的外销瓷，唐代以梅县窑的青瓷为主；五代时除了梅县窑青瓷，还有唐三彩、越窑青瓷、邢窑白瓷、长沙窑瓷器；宋、元时期出口陶瓷的品类繁多，以龙泉窑系青瓷为主，其次是景德镇窑系青白瓷和青花瓷，此外还有磁州窑系、耀州窑系、建窑系瓷器及福建、广东沿海专烧外销瓷的瓷窑产品。进入明清时期，由于世界资本主义商业的大发展，加上明代郑和下西洋促进了中国与海外各国政治、经济和文化的交流，海外对中国瓷器的需求激增。据不完全统计，明崇祯十二年（1639 年），中国运到荷兰的瓷器已达 366000 件。

Chinese porcelain has a long history of being an export product. According to the evidence of archaeological excavation, by the late Tang dynasty, porcelain had appeared as export products in the merchant ships along the Maritime Silk Route. Generally speaking, in the Tang dynasty, the export porcelain of Guangdong province was mainly celadon from Meixian kiln; in the Five dynasties, in addition to the celadon porcelain of from Meixian kiln, there were also Tangsancai (three colours glazed pottery of the Tang dynasty), celadon porcelain of Yue kiln, white porcelain of Xing kiln and porcelain of Changsha kiln; in the Song and Yuan dynasties, there were many kinds of export porcelain wares, most of which were celadon of Longquan kiln, and the secondary were celadon and blue and white porcelain of Jingdezhen kiln, as well as pieces from Cizhou kiln, Yaozhou Kiln and Jian kiln, whereas export porcelain products were especially made in Fujian and Guangdong coastal area. In the Ming and Qing dynasties,

August / 2021.8

4

星期三

Wednesday

辛丑牛年 乙未月

六月廿六

the world's rapid commercial development of capitalism, and Zheng He's trip to Southeast Asia in the Ming dynasty promoted the political, economic and cultural exchanges between China and other countries, which led to a rising overseas demand for Chinese porcelain. According to some incomplete statistics, by the twelfth year of Emperor Chongzhen's reign (1639), 366,000 pieces of porcelain had been transported to the Netherlands from China.

August / 2021.8

5

星期四

Thursday

辛丑牛年 乙未月

六月廿七

《瓷器制运图》插画之一：过岭。这幅插画记录了18世纪清代景德镇外销瓷器以及广州外商的活动。图中装桶的彩瓷由四人担运，而装箩的白瓷只需一人，商人则安坐两人抬的轿中。队伍翻岭后，再循水路沿北江南下，经英德、清远、东莞、佛山到达广州。

An illustration from *The Production and Transportation of Porcelain*: crossing the mountain. The illustrations have depicted the production of export porcelain in Jingde town and the Cantonese Hong merchants' activities. In the picture, the barrels in which there is colourful porcelain needed to be carried by four people, while the barrels in which there is white porcelain can be carried by only one man, and a merchant is sitting inside the sedan carried by two bearers. After crossing the mountain, they would go southward by the waterway of the North River, and then pass Yingde, Qingyuan, Dongguan, and Foshan to arrive at Canton.

外销瓷：中华绝密技 Ⅱ

Export Porcelain: China's Secret Technology Ⅱ

中国瓷器的对外贸易，在清代达到高峰。康、雍、乾三朝瓷器的外销，无论是数量还是质量都远超前朝历代。清代的外销瓷以盘、碗、杯、壶、瓶等日用瓷为主，数量巨大，如 18 世纪初销往欧洲的一套餐具可达 60 到 800 件之多。

为了迎合欧洲人的口味，刺激瓷器出口，增加财政收入，在以广州为中心的广东、福建沿海地区，还出现了特制的外销瓷。在中国传统器形和纹饰之外，诞生了新的瓷器样式：中国器形 + 西方纹饰，以及西方器形 + 西方纹饰。当时西方纹饰题材主要有徽章题材、宗教神话题材、爱情题材和现实生活题材四类。这些中西合璧的瓷器，与其他同样中西交融的外贸商品一起构成了十三行外销鲜明的时代特征。

China's export porcelain trade peaked in the Qing dynasty. The quantity and quality of porcelain exported in the period of Kangxi, Yongzheng and Qianlong were far higher than those of the previous dynasties. In the Qing dynasty, export porcelain wares such as plates, bowls, cups, pots and bottles were mainly made for daily use. At the beginning of the 18th century, a set of table wares sold to Europe might consist of 60 to 800 pieces.

In order to cater to the taste of Europeans and increase the the export volume of porcelain and financial revenue, the customization of export porcelain appeared in Canton, the center in Guangdong and Fujian province. In addition to the traditional Chinese utensils and patterns, new porcelain styles were created: Chinese style shapes with Western patterns, and Western style shapes with Western patterns. At that time, there were four kinds of Western decorative themes: coat of

August / 2021.8

6

星期五

Friday

辛丑牛年 乙未月

六月廿八

arms, religious myths, love stories and daily life. These porcelain wares with Chinese and Western styles, together with other foreign trade commodities marked by Sino-Western features, became the distinctive characteristics of the Canton trade.

清康熙时期英国德克尔家族和沃特金斯家族联姻徽章大盘。盘沿绘有凤凰和花果纹，腹部以卍字锦地纹为地，对称开四小窗以金彩绘花卉纹。内圈金彩饰瓔珞纹一周，盘心绘英国德克尔和沃特金斯家族徽章。（外销瓷中常见两个家族组合徽章的徽章瓷，通常是为家族成员联姻而定制。）

Armorial big plate with gold-painted patterns, reign of Emperor of Kangxi, Qing dynasty. There are Phoenix, flowers and fruits on the rim of the plate. The brocade ground is decorated with Manji patterns, and the four symmetrical small open windows are painted with golden flowers. The inner circle is decorated with Yingluo (keyūra, a kind of necklaces in ancient India) pattern, and the family coat of arms of the Deckers and the Watkins are painted in the center of the plate. (On the export porcelain, it is common to see a combination of two families' coat of arms, which is usually customized for the marriage of family members.)

立秋

Beginning of Autumn

August / 2021.8

7

星期六

Saturday

辛丑牛年 丙申月

六月廿九

清乾隆墨彩描金宗教人物故事图圆碟。

Dish with patterns of Greek mythical figures, reign of Emperor of Qianlong, Qing dynasty.

August / 2021.8

8

星期日

Sunday

辛丑牛年 丙申月

七月初一

推手风车国

The Promoter of Chinese Porcelain

中国陶瓷的海外贸易历史虽然悠久，但直到 17 世纪下半叶才真正大规模进入欧洲市场。在这之前，仅有零星中国瓷器以间接的方式传入欧洲王室。17 世纪，荷兰取代葡萄牙成为海上霸主，接管了葡萄牙人在远东的贸易网络。他们改革了当时的瓷器贸易模式，实行分级制，将瓷器分为三等销往不同国家和地区，其中质量最好的销往欧洲，以完成对欧洲大众市场的普及。

为了更加符合欧洲人的审美，荷兰人向中国工匠提供图纸定制瓷器，其中一种印有贵族家徽的徽章瓷，引发贵族间拿瓷器来进行攀比的风气。此外，荷兰人还造出中国陶瓷的高仿品代尔夫特陶，用来与真正的中国瓷器形成对比，以抬高后者价格。在荷兰人的苦心经营下，17 世纪至 18 世纪中国瓷器在欧洲十分流行，并由此引发了一股中国风潮。

Although the export trade of Chinese porcelain had a long history, it was not until the second half of the 17th century that the porcelain entered the European market on a large scale. Before that, only a few pieces of Chinese porcelain were introduced to the European royal families indirectly. In the 17th century, the Dutch took over the Portuguese's trade network in the Far East. They reformed the porcelain trade mode at that time, implemented the classification system for porcelain, dividing the porcelain into three classes, and sold it to different countries and regions. The porcelain of best quality was sold to Europe, which led to the popularization of porcelain in the European mass market.

In order to be more in line with the European aesthetics, the Dutch provided the Chinese craftsmen with custom-made porcelain drawings, among which there were some with the coat of arms of noble families. So the aristocrats kept up with

August / 2021.8

9

星期一

Monday

辛丑牛年 丙申月

七月初二

each other by showing off their porcelain. In addition, the Dutch also created high-quality copies of Chinese porcelain, Delft pottery, which was compared with the real Chinese porcelain to raise the price of the latter. Under the painstaking management of the Dutch, Chinese porcelain was very popular in Europe in the 17th and 18th century, and led to the fashion of "Chinoiserie".

18世纪初期，荷兰代尔夫特工厂烧制的中国陶瓷高仿品。（图片来源：FOTOE 图片库）

A high-quality copy of Chinese porcelain made by the Delft factory in the Netherlands in the early 18th century. (The picture is bought from the photo gallery FOTOE.)

August / 2021.8

10

星期二

Tuesday

辛丑牛年 丙申月

七月初三

广彩有洋脉

The Western Elements of Guangcai Porcelain

15 世纪中叶，在欧洲出现了一种以珐琅作为颜料，在金、银等金属器物上填绘图案的技术，称为“西洋画珐琅”。所谓“珐琅”是一种玻化物质。它是以长石、石英为主要原料，加入一系列矿物化合物作为助熔剂、乳浊剂和着色剂，经过粉碎、混合、煅烧、熔融、急冷、研磨后得到的颜料。

大约在 17 世纪，画珐琅通过外国商人和传教士传入广州，被广州工匠吸收，结合始于明代的“广州三彩”烧制技术，先创制出“铜胎画珐琅”（也称洋瓷或广州珐琅），后又在白瓷胎上作画，形成“瓷胎画珐琅”（也称珐琅彩瓷），此为广彩的滥觞。

In the middle of the 15th century, in the Flanders area, there appeared a technique called “painted enamel”, using the enamel pigments to fill in the patterns on gold, silver and other metal objects. Enamel is a kind of vitrified material. Feldspar and quartz would be used as its main components. A series of mineral compounds would be added as fluxing agents, opacifiers and colorants, and then the pigments would be crushed, mixed, calcined, melted, quenched and grounded.

In the 17th century or so, the painted enamel technique was introduced to Canton through the foreign merchants and missionaries, and was acquired by Cantonese craftsmen. The combination of the painted enamel technique and the firing technology of “Cantonese three colours” was started in the Ming dynasty, and “the painted enamel on copper" (also known as foreign porcelain or Cantonese enamel) was first created, and followed by “the painted enamel on white porcelain”, (polychrome enamel porcelain), which was the origin of Guangcai porcelain (polychrome porcelain painted in Canton).

August / 2021.8

11

星期三

Wednesday

辛丑牛年 丙申月

七月初四

清中期铜胎画珐琅“功名富贵”纹花口盘。圆盘分八棱，盘心白地彩绘“功名富贵”纹，画面中央描绘了一只公鸡伫立于湖石之上，引颈高鸣。公鸡鸣叫寓意“功名”，配上色彩娇艳的牡丹，取其荣华富贵之意，故为“功名富贵”。这件作品色彩鲜艳，绘工细腻，寓意吉祥，其器形可上溯到 18 世纪初的银托盘，一些托盘可分六、八、十或更多棱。（原件藏于广州十三行博物馆）

Enamel plate with wavy rims and patterns symbolizing merit and fame, mid-Qing dynasty. The round plate has eight arching edges. The center of the plate is a colourful painting symbolizing "merit and wealth". In the middle of the painting, a rooster stands on the rock of a lake, stretching his neck and singing loudly. The pronunciation of "the singing of a rooster" is abbreviated as the pronunciation of "merit", and it is matched with the delicate and colourful peonies, which symbolize prosperity, to form the connotation of "merit and wealth". This work is intense in colours, delicate in carving technique, with an auspicious meaning. The shape of this ware can be traced back to the silver tray made in the early 18th century. Some trays have six, eight, ten or more arching edges. (The original piece is collected in Guangzhou Thirteen Hongs Museum.)

August / 2021.8

12

星期四

Thursday

辛丑牛年 丙申月

七月初五

浮梁造广彩

Guangcai Porcelain

广彩是广州地区釉上彩瓷艺术的简称，是以低温釉上彩的装饰技法，在白色瓷器上用金粉及五彩颜料彩绘，以700℃至750℃低温烧制而成，因构图紧密、炫彩华丽、金光闪闪，宛如华美富丽的锦缎，故而得名。

广彩的出现源于外销需求。进入17世纪，瓷器成为最受外国人欢迎的中国商品之一，清代一口通商时期，不少外国商人依照国外审美对瓷器进行来图定制。为了满足海外市场需求，十三行商人便在景德镇事先烧好素胎白瓷，运到广州后，另觅工匠仿照西洋画法加以彩绘，并且在广州珠江以南一带开炉烘染，制成彩瓷。广彩起于康雍年间，盛于乾嘉时期并流传至今，是清代广州匠人结合西方工艺和审美创制的瓷器新品种。

"Guangcai" is the abbreviation of "polychrome overglaze porcelain painted in Canton". The white porcelain would be painted with gold powder and colourful pigments, and then fired in low temperature of 700 to 750 degrees. It gets this name because of its full composition, fantastic colours, resplendent luxuriousness which looks like gorgeous brocade.

The emergence of Guangcai porcelain was due to the export demand. From the 17th century onward, porcelain became one of the most popular Chinese commodities among foreigners. During the one-port trade period of the Qing dynasty, many foreign merchants customized porcelain according to their aesthetics. In order to meet the demand of overseas markets, the Hong merchants ordered the white porcelain to be fired in Jingde town in advance, and transported it to Canton, where craftsmen would paint it with the Western painting method. After being fired in the kilns on the southern bank of the Pearl River in Canton, the

August / 2021.8

13

星期五

Friday

辛丑牛年 丙申月

七月初六

polychrome porcelain was made. Originated in the reign of Emperors Kangxi and Yongzheng, and flourished in the reign of Emperors Qianlong and Jiaqing, Guangcai porcelain, a new type of porcelain with Sino-Western techniques and aesthetics made by the Cantonese artisans in the Qing dynasty, has been handed down to the present.

August / 2021.8

14

星期六

Saturday

辛丑牛年 丙申月

七月初七

七夕节

Chinese Valentine's Day

清中期铜胎画珐琅山水纹提梁壶。壶身四面开光，开光内绘山水图景，画题为“唐宋八大家”之一的欧阳修《远山》诗句；边饰墨彩描金、洋红卷草纹；盖钮、口沿露铜胎。此壶胎轻体薄，开光绘画的题材呼应诗中悠然自得的意境。这种造型的提壶可上溯到清早期的瓷器，在欧洲，这种提壶多用以盛茶。

Enamel teapot with transom-shaped handle and landscape patterns, mid-Qing dynasty. There are four open windows painted on the pot, and the landscapes are painted inside. A sentence from the poem *Distant Mountain* written by Ouyang Xiu, one of the eight great literati in Tang and Song dynasties, is taken to be title of the painting. The rim is decorated with inky and golden depictions and magenta rolling grasses; the copper of the mouth is exposed. The pot is light and thin in texture, while the paintings inside the open windows echo with the leisure atmosphere of the poem. The teapots made in such shapes can be traced back to the porcelain made in the early Qing dynasty. In Europe, the teapots of this kind are mostly used for storing tea.

19 世纪后期铜胎画珐琅花卉纹瓜形小盖盒。

Enamel copper small pumpkin-shaped covered box with patterns of flowers and branches, late 19th century.

August / 2021.8

15

星期日

Sunday

辛丑牛年 丙申月

七月初八

广州画珐琅

Cantonese Painted Enamel Wares

西洋画珐琅传入中国后，根据制作工艺大致分为广州珐琅（画珐琅）和掐丝珐琅（景泰蓝）两大类。

广州珐琅指在器物铜胎上烧制白釉，再在白釉上以珐琅作画，经多次烧制完成的釉上彩技术，整个过程分五个步骤：铜坯（以铜板打制器形）、挂瓷（在铜坯上施白色珐琅釉，入窑烧制）、绘画（以单色或多色珐琅釉料绘制图案）、烧彩（入窑烧制显色）、镀金（将外露铜边镀金）。广州珐琅内容多以肖像、人物、风景、静物与历史故事、宗教故事为主，风格淡雅、古朴、细致，釉质温润细腻，除了制作瓶、盒、盘、碗之外，还用于制作家具、钟表、化妆品盒子的嵌件，曾经在大众生活中十分普遍的搪瓷制品也是广州珐琅的一种。

After the introduction of Western painted enamel technique into China, the enamel wares could be roughly divided into two types: Cantonese enamel (painted enamel) and filigree enamel (cloisonné) according to the production process.

Cantonese enamel belongs to the overglaze technique. The copper object would be fired after it is painted with a layer of white glaze. Then the patterns would be painted on the white glaze and the ware needs to be fired several times. The whole process is divided into five steps: making a copper ware (by hammering the copper), making the porcelain surface (by applying white enamel on the copper object which would be fired in the kiln), painting (patterns with one or several enamel pigments), firing (so that the colour could be fixed), gilding (the exposed copper edge). The themes of Cantonese enamel are usually portraits, figures, landscapes, still life, historical myths and religious stories. The style is elegant, simple and delicate, and the texture is smooth and exquisite. In addition to making

August / 2021.8

16

星期一

Monday

辛丑牛年 丙申月

七月初九

bottles, boxes, plates and bowls, it is also used as the inlaid objects of furniture, clocks and cosmetic cases. Tangci, a product once very common in public life, is also a kind of Cantonese enamel.

清乾隆款画珐琅果蝶烟壶。清康熙年间，欧洲的西洋画珐琅工艺传入中国，经过中国工匠的实验和改造，形成具有东方特色的画珐琅艺术，至乾隆时期已发展得精密繁复、雍容华贵。（图片采自《粤海珍萃》）

Enamel snuff bottle, reign of Emperor of Qianlong, Qing dynasty. During the reign of Emperor of Kangxi of the Qing dynasty, the painted enamel technique was introduced from Europe into China. After the Chinese craftsmen's experiment and transformation, the art of painted enamel with oriental characteristics was formed. By the period of Emperor Qianlong, it had developed into a sophisticated and elegant artwork. (The picture is taken from *Treasures from Guangdong*.)

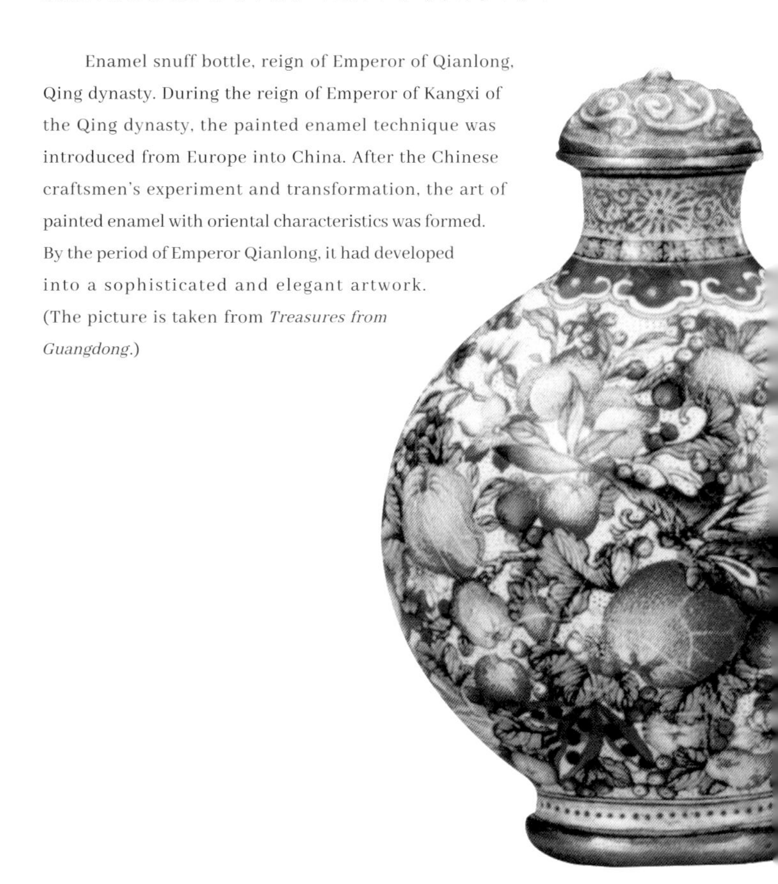

August / 2021.8

17

星期二

Tuesday

辛丑牛年 丙申月

七月初十

灵思立广彩

Lingsi Hall, the Guild of Guangcai Porcelain

“西洋画珐琅”被本土化为“瓷胎画珐琅”后，其制作工艺随优秀的广州匠人传入清廷造办处，烧制专供皇室使用的珐琅彩瓷。这类珐琅彩瓷雍容华贵，集中西风韵于一身，尽显皇家身份的高贵气派，是釉上彩瓷中最为精美的珍品。而更多留在广州的工匠则在大批量生产外销珐琅彩瓷的过程中渐渐形成产品特色。乾隆四十三年（1778年），一个新的行会组织——“灵思堂”出现，标志着广彩作为独立行业的开始。

After the “western painted enamel” was localized into the “painted enamel on porcelain”, some skillful Cantonese artisans introduced it into the royal workshop of inner Qing court, providing the royal family with specialized enamel wares, which were luxurious and expressive with a Sino-Western style, demonstrating their nobility. They were the most exquisite pieces among all the enamel wares. Whereas the export enamel wares made by the craftsmen who stayed in Canton gradually formed their unique characteristics in the process of mass production. In the thirty-fourth year of Emperor Qianlong's reign (1778), a new guild “Lingsi Hall” was set up, which was the beginning of the production of Guangcai porcelain as an independent industry.

August / 2021.8

18

星期三

Wednesday

辛丑牛年 丙申月

七月十一

19世纪铜胎画珐琅花卉纹爵杯。（原件藏于广州十三行博物馆）

Enamel copper goblet with patterns of flowers, 19th century. (The original piece is collected in Guangzhou Thirteen Hongs Museum.)

August / 2021.8

19

星期四

Thursday

辛丑牛年 丙申月

七月十二

模仿西洋风

Imitating the Western Style

康熙、雍正时期为广彩初创期，其彩绘以国画技法为主，没有明显的风格，与景德镇彩瓷区别不大。乾隆时期为繁盛期，引入外国颜料为主材料，在透视、晕染和明暗对比方面则吸收了西洋画技法，绘制内容多半是具有岭南风格的图案，最常用的构图是用花边图案围出若干形状各异的空格，在空格内绘以花卉、物景和人物。也有不划区域，直接以满花彩绘，器形方面则出现大量模仿西洋器皿的款式。

The period of Emperor Kangxi and Emperor Yongzheng was the initial stage of Guangcai porcelain. The painting, mainly based on traditional Chinese painting techniques, without a conspicuous style, was similar to that of Jingde town's porcelain. The period of Emperor Qianlong was its prime when foreign pigments were introduced to be used as the main materials, and Western painting techniques such as perspective, sfumato and chiaroscuro were applied. Most of the contents were patterns with Lingnan style. The most commonly used composition was to enclose a number of spaces of different shapes with floral borders. Flowers, scenery and figures were painted in the spaces. There were also cases in which the surface would not be divided but fully covered with floral patterns. In terms of shapes, there appeared a lot of copies of Western utensils.

August / 2021.8

20

星期五

Friday

辛丑牛年 丙申月

七月十三

August / 2021.8

21

星期六

Saturday

辛丑牛年 丙申月

七月十四

清康熙青花矾红描金花卉纹单耳带盖壶。

Blue and white covered pot with patterns of flowers, reign of Emperor of Kangxi, Qing dynasty.

August / 2021.8

22

星期日

Sunday

辛丑牛年 丙申月

七月十五

清雍正广彩人物纹高足盖碗。

High-footed bowl with patterns of figures, Yongzheng period, Qing dynasty.

织金成广彩

Brocade Ground with Golden Patterns

嘉庆至道光中期是广彩的转型期。此时，中国瓷器在海外市场的热度已渐趋消退，反映在瓷器生产上是彩绘技法回归中国传统勾勒法，施彩颜色种类减少，以大面积使用浓艳的红、绿彩为特征。

道光后期至今是广彩的稳定发展期。进入 19 世纪，广彩出口量已经很低，以内销为主。图案方面，岭南画派以画入瓷，为广彩注入了新的生机。同时，19 世纪初，广彩将锦缎纹样绘制在瓷器上作为装饰，称为“织地”；后普遍用金粉绘制织地，称为“织金地”。“织金地”后来成为广彩最具代表性的风格——堆金积玉。

From the reign of Emperor Jiaqing to the middle period of Emperor Daoguang was the transitional period of Guangcai porcelain. At that time, the gradually faded popularity of Chinese porcelain in the overseas market was reflected in the porcelain production. The painting techniques returned to the traditional Chinese sketching method and the variety of colours was reduced. The porcelain was characterized by the use of intense red and green pigments on a large scale.

From the later period of Emperor Daoguang to the present is the stable developing period of Guangcai porcelain. From the 19th century onwards, the export volume of Guangcai porcelain had become very low, and instead, it was mainly sold in domestic market. In terms of patterns, artists of the Lingnan Painting School painted on the porcelain, bringing new vitality to Guangcai porcelain. Meanwhile, in the early 19th century, the brocade patterns were painted on porcelain as decoration, which was called “weaving the brocade ground”; later on, it was generally

处暑

The Limit of Heat

August / 2021.8

23

星期一

Monday

辛丑牛年 丙申月

七月十六

painted with gold powder, and thus known as "weaving the brocade ground with gold", which later became the most representative style of Guangcai porcelain——heaping up the gold and jade.

清道光广彩花蝶人物纹奖杯形带盖瓶。（原件藏于广州十三行博物馆）

Trophy-shaped vase with patterns of flowers, butterflies and figures, reign of Emperor of Daoguang, Qing dynasty. (The original piece is collected in Guangzhou Thirteen Hongs Museum.)

August / 2021.8

24

星期二

Tuesday

辛丑牛年 丙申月

七月十七

尊贵徽章瓷

The Armorial Porcelain

徽章瓷是广州外销瓷中最特别的一种，具有强烈的标识性和排他性。当时，欧美国家的皇室、贵族、社团、城市、军队首领、商人、政界人士、知识分子和神职人员等把独属的徽章图案烧制在瓷器上，作为自己独有标志。

徽章出现于12世纪，发展至今已成为一套具有世袭性质的个人识别体系，用来显示血统、婚姻、权力和社会地位等。早在16世纪上半叶，中国外销瓷中已经出现定制徽章瓷。到了18世纪，定制徽章餐具在欧洲贵族中形成风尚。由于航路遥远，加上清政府严厉的通商制度，一套徽章瓷从下定到最后收货至少需要一年时间，耗工费时，成本高昂。为了降低买卖双方的成本，徽章瓷大多成套定制，每套数量不下200件。

Armorial porcelain is one of the most special export porcelain in Canton, with strong identity and exclusiveness. At that time, the royal families, aristocrats, societies, cities, military leaders, merchants, politicians, intellectuals and clergies of European and American countries required their coat of arms to be painted on porcelain as their unique symbols.

The coat of arms appeared in the 12th century, and now heraldry has become a system which is used for the hereditary personal identification to show bloodline, marriage, power and social status. As early as the first half of the 16th century, custom-made armorial porcelain already appeared in China's export porcelain. In the 18th century, custom-made table wares with coat of arms became popular among European aristocrats. Due to the long route and the strict trading system of the Qing government, it took at least one year for a set of armorial porcelain to be

August / 2021.8

25

星期三

Wednesday

辛丑牛年 丙申月

七月十八

received after it was ordered. The production of the porcelain was time-consuming and costly. In order to reduce the buyers and sellers' cost, most of the armorial porcelain was customized in sets of no less than 200 pieces.

August / 2021.8

26

星期四

Thursday

辛丑牛年 丙申月

七月十九

清乾隆广彩描金开光山水徽章碟。碟心绘英国汉门德家族徽章，徽章的盔饰为鹰头及海军冠冕，盾牌中的掌形图案为男爵标志。1783 年，安德鲁·汉门德被授予男爵称号，1794 年至 1806 年曾任英国海军总审计官。（原件藏于广州十三行博物馆）

Armorial dish with gold-painted landscape patterns, reign of Emperor of Qianlong, Qing dynasty. In the center of the dish is the coat of arms of the Hamond family in England. The crest of the arms is an eagle's head and a naval crown, and the palm-shaped pattern on the shield is the baron's mark. In 1783, Andrew Hamond was awarded the title of Baron. From 1794 to 1806, he was the general controller of the British navy. (The original piece is collected in Guangzhou Thirteen Hongs Museum.)

洛克菲勒瓷：广彩高精尖

Rockefeller Pattern Porcelain: The High-End Products of Guangcai Porcelain

洛克菲勒瓷不仅指美国洛克菲勒家族购藏的广彩精品，也指一种特别的广彩纹饰。这种纹饰源自英国的一家名为“Lowestoft”的瓷器工厂在18世纪为欧洲贵族设计的纹饰。乾隆晚期，广州工匠把这种纹饰吸收为广彩纹饰的一种。

洛克菲勒瓷讲究用料精纯，纹饰精细，图案结构大体分为中心主题图案（多为珐琅彩绘制的中式庭院人物故事纹）、内边饰（多为茄色织鲨鱼皮锦地）、外边饰（金彩织洋莲锦地）和开光（内圈开光以花鸟纹为主，外圈开光多为麻色或红色中式山水），整体图案金光灿灿，极富立体感和装饰性。洛克菲勒瓷仅生产了20多年，嘉庆中期后便不再生产。

Rockefeller Pattern Porcelain not only refers to the fine porcelain purchased and collected by the Rockefeller family in the United States, but also to a special style of Guangcai porcelain. Originated from a porcelain factory “Lowestoft” in England, this style was designed for the European aristocrats in the 18th century. In the late period of Emperor Qianlong's reign, Cantonese craftsmen applied this style onto Guangcai porcelain.

Rockefeller Pattern Porcelain is made of fine material and decorated by delicate patterns. The patterns generally include central theme patterns (most of which are Chinese gardens, figures and stories painted by enamel), interior decorations (most of which are aubergine shark-skin brocade ground), exterior decorations (brocade ground with golden lotuses) and open windows (inner windows mainly filled with flowers and birds, while outer windows mainly filled with Chinese landscapes painted in linen and red). The overall patterns are brilliant and splendid, highly realistic and

August / 2021.8

27

星期五

Friday

辛丑牛年 丙申月

七月二十

decorative. Rockefeller porcelain had been produced for more than 20 years, and the production stopped after the middle period of Emperor Jiaqing's reign.

清同治广彩人物纹镂空果篮托碟。

Pierced oval fruit tray with patterns of figures, reign of Emperor of Tongzhi, Qing dynasty.

August / 2021.8

28

星期六

Saturday

辛丑牛年 丙申月

七月廿一

清光绪广彩花鸟人物纹带托盖碗。

Covered bowl with patterns of figures, flowers and birds, reign of Emperor of Guangxu, Qing dynasty.

August / 2021.8

29

星期日

Sunday

辛丑牛年 丙申月

七月廿二

“满大人”：中国合家欢

“Mandarins”: the Happy Chinese Families

“满大人”一词最早出现于明代，是英文中对中国乃至整个亚洲古代官吏的泛称。自康熙时起至道光，在广彩、粉彩外销瓷中出现了一种具有时代特征的清装人物纹样，也被称为“满大人”。

“满大人”纹样多出现于外销瓷，在内销瓷里很少见，是商人们为满足西方消费者对中国式生活的好奇而让中国工匠设计特制的。“满大人”纹饰多以两代或三代同堂的“合家欢”为主题，人物被安置在中式庭院、厅堂或回廊之中，悠然自得地休闲娱乐，背景是中式山水、田园、池塘、亭台、楼阁等，是清代中国社会富裕阶层日常生活的理想化描绘，为西方人了解当时的中国文化提供了一个窗口。

The term “Mandarin”, first appeared in the Ming dynasty, was a general term in English referring to the officials in ancient China or even the whole Asia. From the time of Emperor Kangxi to Emperor Daoguang, the figures dressed in the Qing costumes with features of that time appeared on the export porcelain such as Guangcai and Famille-rose porcelain were also called as “Mandarin”.

The patterns of “Mandarins” mostly appear on the export porcelain, and are rarely seen on the domestic porcelain. The patterns were specially designed by Chinese craftsmen to satisfy the Western consumers' curiosity about Chinese life. The figures of “Mandarins” are mostly found in the themes of “family union” where two or three generations gather together. The characters are in Chinese courtyards, halls or cloisters for leisure and entertainment. At the background there are Chinese landscape, gardens, ponds, houses, pavilions and so on. They are the ideal depictions of daily life of the Chinese rich class in the Qing dynasty, providing westerner with a glimpse into the Chinese culture of that time.

August / 2021.8

30

星期一

Monday

辛丑牛年 丙申月

七月廿三

August / 2021.8.8

31

星期二

Tuesday

辛丑牛年 丙申月

七月廿四

清代木雕金漆“满大人”像。“满大人”原是西方人对中国清代官员的称呼，后也将绘有清代人物服饰纹的瓷器或雕像器物称为“满大人”。（原件藏于广州十三行博物馆）

Mandarins, wooden lacquer sculptures made in the Qing dynasty. Originally, the Westerners called the Chinese Qing officials as “Mandarins”, and later on, the porcelain and sculptures with the Qing figures, were also called “Mandarins”. (The original piece is collected in Guangzhou Thirteen Hongs Museum.)

September

一	二	三	四	五	六	日
		1 廿五	2 廿六	3 廿七	4 廿八	5 廿九
6 三十	7 白露	8 初二	9 初三	10 教师节	11 初五	12 初六
13 初七	14 初八	15 初九	16 初十	17 十一	18 十二	19 十三
20 十四	21 中秋节	22 十六	23 秋分	24 十八	25 十九	26 二十
27 廿一	28 廿二	29 廿三	30 廿四			

“伊万里”外销瓷：东洋瓷器中国造 Ⅰ

Imari: Export Porcelain of Japanese Style Made in China Ⅰ

明代万历年间，日本在文禄庆长之战后，从朝鲜半岛带回大量陶工，开始了日本的瓷器烧制。他们以中国景德镇青花瓷为样本，将中国和朝鲜的制瓷技术结合起来，烧出了“伊万里”瓷器。

“伊万里”瓷器装饰在布局上完全继承中国景德镇的装饰模式，即主纹饰与辅纹饰结合，器身通体装饰或开光主纹饰、辅纹饰与边饰装饰带结合。1690年之后，又逐渐形成了在釉下青花的基础上施加釉上彩装饰，最后再以金粉加彩点缀的做法，令瓷器整体风格显得绚丽夺目、雍容华贵，正迎合了当时欧洲所盛行的洛可可艺术风格。

In the period of Emperor Wanli's reign of the Ming dynasty, the Japanese brought back a large number of potters from the Korean Peninsula after the Bunroku-keichou Campaigns, and began to produce porcelain in Japan. They took the blue and white porcelain of Jingde town as the sample, combined the Chinese technology of making porcelain with that of Korea, and produced the “Imari” porcelain.

In terms of composition and decoration, the Imari porcelain completely inherits the decorating style of Jingde town's porcelain in China, combing the main patterns with the auxiliary ones, the decoration all over the porcelain or the open windows of the main theme with the auxiliary patterns and the border patterns. After 1690, overglaze decoration was applied onto the underglaze blue and white porcelain which would be finally decorated with golden powder with colourful ornament, so that the overall style of the porcelain appeared to be gorgeous, eye-catching, splendid and luxurious, and catered to the Rococo style prevailing in European art at that time.

September / 2021.9

1

星期三

Wednesday

辛丑牛年 丙申月

七月廿五

清雍正墨彩描金天主教故事图碟。（原件藏于广州十三行博物馆）

Chinese Imari porcelain: dish with inky and golden illustration of Catholic story, Yongzheng period, Qing dynasty. (The original piece is collected in Guangzhou Thirteen Hongs Museum.)

September / 2021.9

2

星期四

Thursday

辛丑牛年 丙申月

七月廿六

“伊万里”外销瓷：东洋瓷器中国造 Ⅱ

“Imari”: Export Porcelain of Japanese Style Made in China Ⅱ

“伊万里”瓷器在日本发展最快的时期，正处于清初海禁实施期。急于找到中国瓷器替代品的荷兰商人发现日本青花瓷、白瓷的烧制技术已经十分成熟，转而在日本订购“伊万里”瓷器销往欧洲，很快就占领了欧洲市场。1684年以后，随着海禁政策的解除，中国逐步恢复对外贸易。为了快速抢回欧洲市场，景德镇仿造“伊万里”瓷器的样式，也烧制青花矾红描金瓷器，称为“中国伊万里”。和日本瓷器相比，中国青花矾红描金瓷胎质细腻，青花发色较淡雅，画工洒脱、不拘束，很快便重新夺回欧洲市场。1757年，荷兰东印度公司正式停止购买日本瓷器，“伊万里”瓷器向欧洲的出口正式落下帷幕。

Imari porcelain appeared in the period of Japan's most rapid development, during which the maritime affairs was banned in the early Qing dynasty. Dutch traders were eager to find a substitute for Chinese porcelain and found that the firing technology of Japanese blue and white porcelain was very mature. The Imari porcelain was ordered by the Dutch in Japan, sold to Europe, and soon occupied the European market. After 1684, with the lifting of the ban on the maritime affairs, China gradually resumed foreign trade. In order to get back the European market quickly, artisans of the Jingde town imitated the Imari porcelain's style and at the same time produced blue and white porcelain with alum-red and golden patterns, which was called “Chinese Imari”. Compared with the Japanese porcelain, China's porcelain with alum-red and gold patterns was more delicate. Its blue colour was light and elegant, and the painting style was free and uninhibited, so it soon regained the European market. In 1757, the Dutch East India Company officially stopped buying porcelain from Japan, and the export of Imari porcelain to Europe officially came to an end.

September / 2021.9

3

星期五

Friday

辛丑牛年 丙申月

七月廿七

清乾隆广彩青花人物纹狮钮盖瓶。

Blue and white vase with figural patterns and lion-shaped knob, reign of Emperor of Qianlong, Qing dynasty.

September / 2021.9

4

星期六

Saturday

辛丑牛年 丙申月

七月廿八

September / 2021.9

5

星期日

Sunday

辛丑牛年 丙申月

七月廿九

清乾隆广彩人物纹花口扁瓶。

Vase with patterns of figures, reign of Emperor of Qianlong, Qing dynasty.

广彩世无双 I

The Peers of Guangcai Porcelain I

广彩、粉彩与五彩同属于中国传统的釉上彩瓷，但在图案、彩料和技法上又各有特点。粉彩和五彩以内销为主，因此图案也以符合中国人审美的传统题材为主；广彩在传统题材之外，还有专为外国主顾定制的西洋器形和图案。广彩所使用的颜料多半由广州艺人自制或改配外国颜料，颜色有西红、干大红、茄紫、广翠、麻色、粉绿、金彩、鹤春等；五彩则使用中国传统颜料，俗称“古彩”，基本色调为红、黄、绿、蓝、黑等，成色浓艳而不刺目；粉彩颜料多为进口货，颜色上与广彩接近，但金彩、赭色使用相对较少。

Guangcai, Famille-rose and Five-coloured porcelain are all Chinese traditional overglaze porcelain, but they have their own characteristics in terms of patterns, materials and techniques. Famille-rose and Five-couloured porcelain were mainly sold in China, so the patterns were also based on traditional themes in accordance with Chinese aesthetics; in the case of Guangcai porcelain, besides traditional themes, there were also Western shapes and patterns specially designed for foreign customers. Most of the pigments used in Guangcai porcelain were made by Cantonese artists or adopted from foreign pigments, including watermelon red, Gandahong (a kind of Chinese red colour), aubergine, colour of Cantonese-jade, colour of linen, light green, gold, greenish blue, etc.; the Five-coloured porcelain, painted with traditional Chinese pigments, commonly known as the “ancient colours”, whose basic colours were red, yellow, green, blue, black, and so on, was rich in color but not dazzling; in terms of Famille-rose, most of the pigments were imported, similar to those of Guangcai porcelain, but comparatively speaking, gold and ochre were used less.

September / 2021.9

6

星期一

Monday

辛丑牛年 丙申月

七月三十

清光绪五彩人物纹大盘。（原件藏于广州十三行博物馆）

Polychrome big plate with figural patterns, reign of Emperor of Guangxu, Guangxu period Qing dynasty. (The original piece is collected in Guangzhou Thirteen Hongs Museum.)

白露

White Dew

September / 2021.9

7

星期二

Tuesday

辛丑牛年 丁酉月

八月初一

广彩世无双 Ⅱ

The Peers of Guangcai Porcelain Ⅱ

广彩的绘画技法兼容并蓄，吸收了五彩、粉彩、珐琅彩和西洋技法，前期以没骨法为主进行彩料渲染，后期则受岭南画派影响逐渐采用工笔画法。五彩也以没骨法为主，采用单线平涂上色，红绿彩十分浓厚，看上去质感坚硬，故又称“硬彩”。粉彩则先用含砷的“玻璃白”打底，在色料中掺入粉质，以芸香油调色，采用黑线勾勒边线，再渲染彩料，这样绘制而成的图案色彩粉润柔和，所以又被称为“软彩”。由于粉彩颜料中掺有铅粉并加施玻璃白，粉彩具有施彩厚、花纹凸起、层次分明、立体感强的特点。

The painting techniques of Guangcai porcelain were inclusive, absorbing techniques from Five-coloured porcelain, Famille-rose, enamel wares and western paintings. In the early period, artisans mainly used boneless brushstrokes to render the colours, and in the later period, influenced by Lingnan Painting School, they gradually adopted meticulous brushstrokes. The Five-coloured porcelain was also mainly painted with boneless brushstrokes. The artisans drew the contours with lines and applied the colours with flat strokes. The red and green colors are extremely thick and dense, and the texture looks solid so that it is called “hard colours”. In the case of Famille-rose, the “glassy white” containing arsenic was applied on the base, and the pigments would be added with powder and toned with rutaceae oil. The artisans drew the contours with black lines, and then rendered the colours. In this way, the colours of the patterns are soft and tender, so the Famille-rose is also called “soft colours”. Because the pigments are mixed with lead powder and powder of glassy white, the Famille-rose porcelain is characterized by the thick colours, protruding patterns, distinct gradation and a strong three-dimensional effect.

September / 2021.9

8

星期三

Wednesday

辛丑牛年 丁酉月

八月初二

清乾隆粉彩堆塑松鼠葡萄纹杯。（原件藏于广州十三行博物馆）

Cup with protruding patterns of grapes, reign of Emperor of Qianlong, Qing dynasty. (The original piece is collected in Guangzhou Thirteen Hongs Museum.)

September / 2021.9

9

星期四

Thursday

辛丑牛年 丁酉月

八月初三

丝绸耀东方

Silk from the East

中国是世界上最早养蚕缫丝织布的国家，丝织品自古便是中国对外贸易中长盛不衰的商品，也承载了西方对东方最早的想象，海陆丝绸之路更是因其得名。清代十三行时期，外销的丝织品主要有绸缎布匹、生丝原料和手帕、手袋等丝绸产品，数量巨大，如乾隆时期每年外销的丝织品达二三十万斤，仅次于茶叶出口量，推动了以广州为中心的岭南丝织业飞速发展，不仅成立了行业组织——锦纶会馆，还革新织造技术，发展出广纱、广缎、絽(xié)、五丝、八丝、香云纱等名品，光华灿然，不褪色，不沾尘，不易皱，故时人有“广纱甲天下，絽次之”的说法。

China is the first country in the world to raise silkworms to produce silk. Silk has always been a favourable commodity in China's foreign trade since ancient times. It also aroused the Westerner's earliest imagination towards the East. The Overland and Maritime Silk Routes were named after it. During the Canton trade period in the Qing dynasty, the silk products exported in large amount were silk fabrics, raw silk, handkerchiefs, handbags and other categories. During the reign of Emperor Qianlong, the annual export volume of silk products reached 200,000 to 300,000 catties, second to the export volume of tea. This promoted the rapid development of silk industry in Lingnan region with Canton as the center. Not only the industrial organization——Hall of Brocade was established, but the weaving technology was also innovated to produce famous products such as Cantonese gauze, Cantonese fabric, Xie, Wusi, Basi and Xiangyun gauze, which were bright, colorfast, dust-free and wrinkle-free. Therefore, there was a saying at that time "Cantonese silk was the best in the world".

September / 2021.9

10

星期五

Friday

辛丑牛年 丁酉月

八月初四

教师节

Teacher's Day

清末红缎地广绣金线绣花卉纹对襟汉式氅衣。氅衣为清代妇女服饰，一般穿在衬衣外面。这件氅衣以红色素缎为面料，全身绣四季花卉，包括莲花、梅花、芙蓉、牡丹等，排金花卉富丽堂皇。袖口饰鹅黄色菱格暗花绸绣百花纹挽袖，衣边及领口饰鹅黄色绸绣花蝶纹、卍字纹，钉金绣缠枝纹、白地碎花窄绦，与面料图案相呼应。绣工平齐细密，运用了打籽针、捆咬针法，变化娴熟。

Canton embroidery surcoat of Han dynasty style with golden floral patterns on red satin, late Qing dynasty. The surcoats, usually put outside the shirts, were women's clothes in Qing dynasty. This surcoat is mainly made of red satin and embroidered with flowers of four seasons, including lotuses, plum blossoms, hibiscus, peonies, etc. The embroidered golden floral patterns are magnificent. The sleeves made of light yellow silk are decorated with undertint rhombic patterns and floral patterns of all kinds; while the light yellow hem and neckline are decorated with patterns of butterflies, Manji patterns, golden patterns of twigs as well as white braids of small floral patterns, which are matched with the satin, the main fabric of the surcoat. The embroidery is neat and delicate. The embroiderers were adept at variations of "Dazi" and "Kunyao" needling techniques.

September / 2021.9

11

星期六

Saturday

辛丑牛年 丁酉月

八月初五

民国白缎地广绣百鸟纹对襟长氅衣。氅衣以白色素缎为面料，前后满绣凤凰、仙鹤、老鹰、锦鸡、绶带鸟、鸳鸯、鹦鹉、翠鸟、猫头鹰等多种鸟类，以及荷花、梅花、雏菊等数种花卉，下摆绣了大面积的高低错落的水波纹，构图饱满，配色富丽。这件氅衣手工精美，制作华丽，寓意吉祥，运用了线绣和绒绣及多种针法，其中小雏菊运用的是打枣针，此外还有金鱼起鳞片、流星赶月凤尾等多种针法表现形式，绣出的花鸟逼真精美，是一件不可多得的广绣服饰精作。

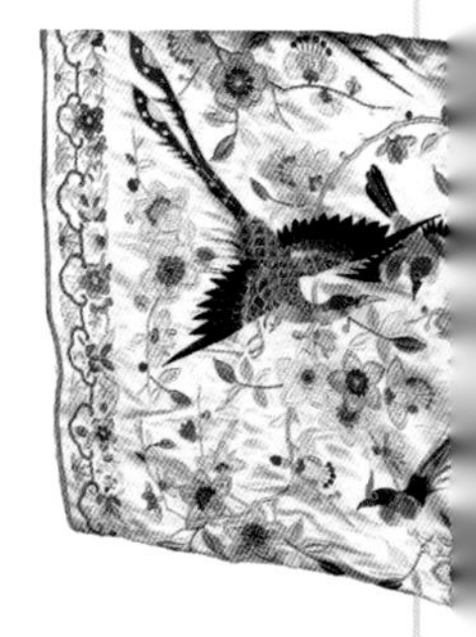

Canton embroidery surcoat with patterns of birds and flowers on white satin, period of the Republic of China. The surcoat is made of white satin. In both the back and front, it is embroidered with patterns of phoenix, crane, eagle, golden pheasant, paradise flycatcher, mandarin ducks, parrot, kingfisher, owl and several kinds of flowers, such as lotuses, plum blossoms and daisies. The lower hem is largely embroidered with patterns of surging water ripples. The composition is full and the colours are rich. This surcoat is exquisite in craftsmanship, gorgeous in production and auspicious in meaning. The embroiderers applied many needling methods, such as thread embroidery, velour embroidery. Among them, the "Dazao" needling technique is used to embroider the little daisies. Besides, there are many other kinds of needling techniques such as "scraping up the goldfish scales" and "decorating the phoenix tail with shinny sequins that look like meteors around the moon". The embroidered flowers and birds are lifelike and exquisite. This surcoat is a rare fine work of Cantonese embroidery.

September / 2021.9

12

星期日

Sunday

辛丑牛年 丁酉月

八月初六

锦纶会馆：丝行老龙头

The Hall of Brocade: the Headquarter of the Textile Industry

始建于雍正元年（1723 年）的锦纶会馆，已经有近 300 年历史了。这座三路三进的祠堂式建筑占地面积约 700 平方米，是广州唯一保留下来的丝织行业东家会馆。锦纶会馆最兴旺的时候，旗下有数百家机户和商号，5000 张以上机张，从业工人达数万。锦纶会馆对生产品质要求很高，被它认可、打上“锦纶堂”名号的丝织品，无一不是织造精美、质量上乘，不仅是外销精品，也常出现在宫廷贡品的名单里。

广州的丝织行业，按产品种类可分为五大行：朝蟒行、十八行、十一行、金彩行、通纱行。每年年初，各行业领袖都集会于锦纶会馆，共同商定丝织商品的规格和价格，基本掌控了十三行丝绸贸易的定价权和话语权。

Founded in the first year of Emperor Yongzheng's reign (1723), Hall of Brocade has a history of nearly 300 years. The ancestral hall building covers an area of about 700 square meters. It is the only remaining merchant guild of silk industry in Canton. When the Hall of Brocade was at its prime, it owned hundreds of factories and trade houses, more than 5,000 machines and tens of thousands of workers. Hall of Brocade set a high standard for production quality. All the silk fabrics recognized by it with a mark “Hall of Brocade” were exquisite and of high quality. They were not only export products, but also often appeared in the list of tributes paid to the court.

According to the types of products: the Cantonese silk industry could be divided into five major categories: Chaomang, Shiba, Shiyi, Jincai and Tongsha. At the beginning of each year, leaders of every category gathered in the Hall of Brocade to discuss the standards and prices of silk products, and they had largely owned the privilege to control the prices of the silk as well as the management in the Canton trade.

September / 2021.9

13

星期一

Monday

辛丑牛年 丁酉月

八月初七

清代广东画师吴俊绘制的纸本水彩画：《牵经》（左）和《丝商》（右），描绘了清代广州丝织女工牵经的情景，以及广州丝商装卸走水路运送的生丝的情景。（图片来源：FOTOE 图片库）

Drawing the Warp Yarns and *The Silk Merchant*, gouache paintings by Wujun, a Guangdong painter in the Qing dynasty. The two paintings depict the scenes: the Cantonese female workers were drawing the warp yarns; the silk merchants had the workers unload the raw silk, which was transported by waterway. (The picture is bought from the photo gallery FOTOE.)

September / 2021.9

14

星期二

Tuesday

辛丑牛年 丁酉月

八月初八

香云纱：纱中软黄金

Xiangyun Gauze: Precious as Gold

香云纱出现于清代道光年间，是广纱中最贵重的品种，当时一匹香云纱大约价值白银12两，故素有“纱中软黄金”之称。香云纱仅在广东佛山地区生产，它与其他丝织品最大的不同之处就是多了“晒莨”“涂泥”两个步骤。

制作香云纱时，先在白色的真丝坯上，涂上一种叫薯莨的野生植物汁液，以及采自珠江三角洲含多种矿物质的河泥，在阳光的暴晒、河水的冲洗、草地的温润下，使二者神奇融合。整个晒莨过程需要7天时间及30多道工序，经过晒莨后的成品称“莨绸”，凉爽宜人，轻薄柔软富有身骨，越洗越柔韧，因穿在身上发出“沙沙”的声音，又被称为“响云纱”，后演变为谐音“香云纱”。

Xiangyun gauze, appeared in the period of Emperor Daoguang's reign of the Qing dynasty, is the most valuable kind of Cantonese silk. At that time, a roll of Xiangyun gauze was worthy of about 12 tael of silver, and known as “the soft gold”. Xiangyun gauze is only produced in Foshan, Guangdong province. The biggest difference between Xiangyun gauze and other silk fabrics is that there are two extra steps, “sunning the silk” and “daubing with the mud”.

When making the Xiangyun gauze, first of all, the white silk will be coated with the juice of a wild plant called dioscorea cirrhosa and the mud contains various minerals in the Pearl River Delta. With exposure under the sun, the water's washing and the warmth of the grassland, the silk and the juice will be magically blended. The whole sunning process takes 7 days with more than 30 processions. After being sunned, the finished product is called “dioscorea cirrhosa silk”. It feels cool and pleasant, and is light, soft and elastic. The more it is washed, the more flexible it will become. Because it makes the rustling sound while wearing it, it is also called as

September / 2021.9

15

星期三

Wednesday

辛丑牛年 丁酉月

八月初九

"Xiangyun gauze" (gauze that makes a rustling sound which can be heard above the clouds), and then it is replaced by the homophone "Xiangyun gauze" (gauze with a scent that can be smelled above the clouds).

September / 2021.9

16

星期四

Thursday

辛丑牛年 丁酉月

八月初十

十三行的如意洋行，专做丝绸生意。（图片来源：FOTOE 图片库）

Ruyi Hong in the trading region Canton is specialized in the silk trade. (The picture is bought from the photo gallery FOTOE.)

广缎：“银钱堆满十三行”

Cantonese Fabric: “Silver Has Been Heaped up in the Trading Region.”

“五丝八丝广缎好，银钱堆满十三行。”清代著名学者屈大均的这首竹枝词，描绘了十三行广缎贸易的盛况。缎指的是一种质地厚密而有光泽的丝织品。清代广东生产的广缎，不仅出口海外，也是享誉全国的名品。

广缎所用原料很讲究，皆为江南地区的吴丝，其织造技法与云锦中的织锦缎很相似，以多彩的纹纬妆彩提花，属重经重纬的高级丝织物，图案以小巧细碎的满地花为主，喜欢采用鲜艳而反差强烈的色彩搭配，如橘黄、玫红、粉红、藕荷、大绿和白色等。“五丝”和“八丝”是广缎中的两个品种，指的是经纬采用两种色彩，经面做地，纬面起花，以五枚、七枚、八枚或不规则缎纹的正反缎形式织成的织物。五枚缎即五丝，八枚缎即八丝。

"Wusi and Basi are fine Cantonese satin; the silver thus has been heaped up in the trading region." This poem was written by Qu Dajun, a famous scholar in the Qing dynasty, and it describes the prosperity of the silk trade in Canton. Satin is a kind of silk with thick and glossy texture. Cantonese satin produced in Guangdong province in the Qing dynasty was not only exported overseas, but also famous throughout the country.

Cantonese satin was made of fine materials, all of which was Wu silk from the southern region of the Yangtze River. With weaving techniques similar to those of brocade-weaving, it was a high-end silk product with colourful floral patterns, emphasizing the warps and wefts. The patterns are mainly small and delicate flowers all over the surface. The colours are bright with sharp contrast, including orange, rose, pink, lotus colour, sharp green and white. “Wusi” and “Basi” are two kinds of Cantonese satins, whose warps and wefts were of two different colours, with the

September / 2021.9

17

星期五

Friday

辛丑牛年 丁酉月

八月十一

warp surface as the ground, the weft surface as the floral patterns, and Wumei, Qimei, Bamei and other irregular forms as the fabric textures. Wumei, namely Wusi, referred to a cluster of silk consisting of five wefts; Bamei, namely Basi, referred to a cluster of silk consisting of eight wefts.

民国红缎地广绣花蝶仙鹤纹对襟汉式氅衣。

Canton embroidery surcoat of Han dynasty style with patterns of flowers, butterflies and cranes on red satin, period of the Republic of China.

18

星期六

Saturday

辛丑牛年 丁酉月

八月十二

清末蓝缎地广绣花蝶纹对襟汉式氅衣。这件对襟氅衣以蓝色素缎为面料，织造精致细密。衣身绣蝴蝶牡丹、折枝花卉纹，牡丹花使用三色丝线和打籽针法。下摆绣莲花海浪纹，全身图案左右对称，构图疏朗，五彩设色，自然秀美。

Canton embroidery surcoat of Han dynasty style with patterns of butterflies on blue satin, late Qing dynasty. This surcoat, made of blue satin, is delicately woven. The main part is embroidered with patterns of butterflies, peonies, twigs and flowers. The embroiderers used the three-color silk threads and “Dazi” needling method to embroider the peonies. The lower hem is embroidered with patterns of lotuses and waves. The patterns on the left and right are symmetrical, and the composition is clear, bright, and colourful with a natural beauty.

September / 2021.9

19

星期日

Sunday

辛丑牛年 丁酉月

八月十三

广绣：英王甄选品

Cantonese Embroidery: the Products Chosen by Queen Elizabeth

广绣是中国四大名绣之一——粤绣的一个分支，分为刺绣字画、刺绣戏服和珠绣等品类，以构图饱满、形象传神、纹理清晰、色泽富丽、针法多样、善于变化的艺术特色而闻名。

明代正德九年（1514 年），一个葡萄牙人在广州购得一块龙袍绣片，回国献给国王后受到重赏，从此广绣成为外销的主要手工艺品。英国女王伊丽莎白一世也是广绣的追捧者，亲自倡导成立英国刺绣同业公会，按广绣作坊形式组织英王室绣庄，从中国进口丝绸和丝线，加工绣制贵族服饰。到了清代，英国商人拿着服饰图样到广州绣坊订货，促使广绣吸收了西洋油画技法，进一步形成了立体透视、光影富于变化的特点。广绣深受西方消费者欢迎，被誉为“中国给西方的礼物”。

Yue Embroidery is one of the four famous embroideries in China. Its branch Cantonese embroidery is divided into several kinds such as embroidery of calligraphy and painting, costumes, embroidery with pearls, etc., famous for its rich compositions, vivid images, clear textures, rich colours, diversified needlework techniques and changeful artistic features.

In the ninth year of Emperor Zhengde's reign (1514) of the Ming dynasty, a Portuguese bought a piece of embroidered dragon robe in Canton. After returning home he offered it to the king, and received a great reward. Since then, Cantonese embroidery became a main handicraft for export. Queen Elizabeth I was also a fan of Cantonese embroidery. She personally advocated the establishment of the British embroidery guild, the organization of the royal embroidery workshop according to the operation of Cantonese embroidery workshop, and the import of silk and silk

September / 2021.9

20

星期一

Monday

辛丑牛年 丁酉月

八月十四

threads from China to process and embroider clothes for the aristocrats. In the Qing dynasty, British merchants came to Cantonese embroidery shops to order clothes to be made according to the patterns provided, triggering Cantonese embroidery to absorb the Western oil painting techniques and further form the characteristics of three-dimensional perspective as well as the change of light and shadow. Cantonese embroidery was very popular with Western consumers and known as "a Chinese gift given to the West".

清代广绣《三羊开泰》(左)及其局部细节(右)。(图片来源：FOTOE图片库)

Sheep as Auspicious Emblem with details, Cantonese embroidery of the Qing dynasty. (The picture is bought from the photo gallery FOTOE.)

September / 2021.9

21

星期二

Tuesday

辛丑牛年 丁酉月

八月十五

中秋节

Mid-Autumn Festival

粤剧戏服：广绣织云霞

Costumes of Cantonese Opera: Beautiful as Colourful Clouds

明代嘉靖、万历年间，粤剧在佛山地区诞生，粤剧戏服也随之发展起来，成为广绣的一个门类，至今已有 300 余年历史。粤剧戏服制作最早便在今越秀区人民南路状元坊附近，鼎盛时期这里聚集了 50 多家刺绣作坊、3000 多个绣娘。

由于舞台表演讲求戏曲效果，要求服饰图案必须醒目，因此粤剧戏服上的刺绣与刺绣字画不同，不以精细著称，更重质感和效果。戏服刺绣除了绒线绣、丝线绣、金银线绣外，还有片绣、珠绣等，形成了粤剧戏服构图饱满、色彩浓艳且对比强烈的特征，时人认为“虽京师歌楼，无其华靡”，因而清末宫廷的皇室戏班，每年都会特意前来订购大量蟒袍玉带、凤冠霞帔等服饰。

During the reign of Emperor Jiajing and Emperor Wanli of the Ming dynasty, Cantonese opera came into being in Foshan area and the costumes of Cantonese opera became a category of Cantonese embroidery. Up till now, it have already had a history of more than 300 years. The costumes of Cantonese opera were first produced near Zhuangyuanfang (the lane of the top student), on South Renmin Road, Yuexiu District. During its heyday, there were more than 50 embroidery workshops and more than 3000 embroiderers.

Due to the emphasis on the stage performance effect, the costume patterns must be eye-catching, so the embroidery patterns on the costumes of Cantonese opera were different from those of the embroidery of calligraphy and painting, not known for delicacy, but more emphasized the texture and effect. In addition to velvet, yarn, gold and silver threads, there were also sequins and pearls added on the embroidery, making the Cantonese opera costumes to be characterized with full composition, rich colours with strong contrast. People thought that “the Cantonese

September / 2021.9

22

星期三

Wednesday

辛丑牛年 丁酉月

八月十六

opera costumes are even more luxurious than those in the opera houses of Beijing". Therefore, the royal troupes in the late Qing court came here to order a large number of robes of python patterns, jade belts, phoenix crests, colourful capes and other costumes every year.

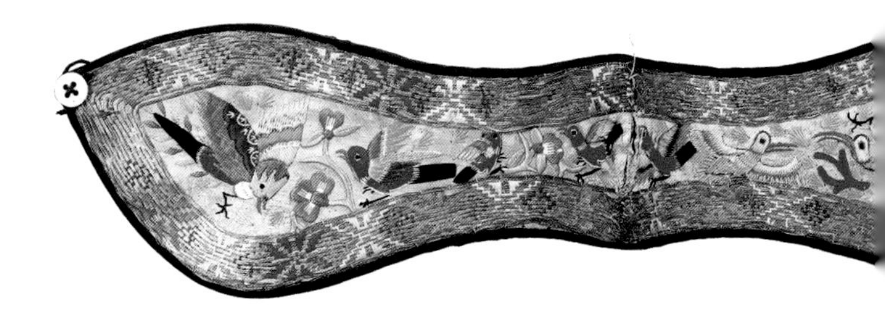

民国广绣梅花喜鹊护额。（原件藏于广州十三行博物馆）

Headband with patterns of magpies and plum blossoms, period of the Republic of China. (The original piece is collected in Guangzhou Thirteen Hongs Museum.)

秋分

Autumnal Equinox

September / 2021.9

23

星期四

Thursday

辛丑牛年 丁酉月

八月十七

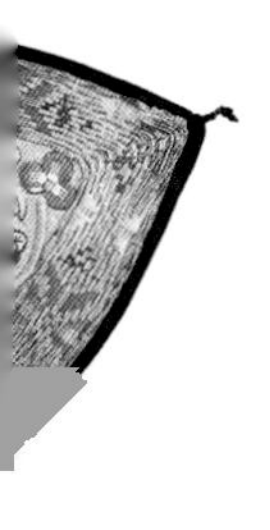

“真金白银”金银绣

Gold and Silver Embroidery

“海纳百川”是岭南文化特征之一，深刻体现于岭南人文的方方面面，驰名中外的广绣同样是兼容并蓄的结果。广绣中的“金银绣”，来自于宁波刺绣中的金银彩绣，指在真丝缎子和其他原料上，用金银线和各色丝线刺绣图案，色彩和谐，古朴雅致。金银绣中的“盘金绣”，则是将金线在缎面上盘出形状，然后用丝线固定而成。盘金绣成品金光闪闪、富丽堂皇，在宁波刺绣里多用来刺绣龙凤图案，据说刺绣一幅精致的龙凤纹盘金绣大约需要消耗一两黄金。其技法被广绣吸收后，题材拓宽至花鸟纹等，图案突出于缎面之上，层次丰富，极有立体感。

“Being open-minded as the sea embracing all the rivers” is one of the characteristics of Lingnan culture, and it is deeply reflected in all aspects of Lingnan humanities. The Cantonese embroidery which is famous at home and abroad is also the result of inclusiveness. The gold and silver embroidery, a category of Cantonese embroidery, originated from the colourful embroidery with golden and silver patterns in Ningbo. On the satin made of fine silk and other materials, the patterns would be embroidered with golden, silver and other colourful threads. The colours were harmonious, vintage and elegant. The technique Panjinxiu (circling golden threads) in “gold and silver embroidery” is to create the shapes with golden threads on a satin, and then fix them with silk threads. The finished products of Panjinxiu technique were glittering and magnificent. This technique was mostly used to embroider dragon and phoenix patterns in Ningbo embroidery. It was said that it took about one or two tael of gold to embroider a piece of delicate Panjinxiu product with dragon and phoenix patterns. After the technique was absorbed into Cantonese embroidery, the

September / 2021.9

24

星期五

Friday

辛丑牛年 丁酉月

八月十八

themes were broadened to include the patterns of flowers and birds. The patterns, protruding from the satin surface, are rich in gradation with a stereoscopic effect.

清末黑绒地金绣团鹤暗八仙纹对襟宗教服。

Religious robe with golden patterns of cranes and eight immortals on black satin, late Qing dynasty.

September / 2021.9

25

星期六

Saturday

辛丑牛年 丁酉月

八月十九

清末黑缎地花卉四品云雁纹氅衣。

Canton embroidery surcoat of the fourth-ranking with patterns of a goose and clouds on black satin, late Qing dynasty.

September / 2021.9

26

星期日

Sunday

辛丑牛年 丁酉月

八月二十

广器：漆品生黄金

Cantonese Lacquer Wares: Precious as Gold

广东漆器最早出现于战国时期，到明清时期达到发展高峰，以潮州金漆木雕、广佛髹漆家具和阳江漆器为代表。清代，广东漆器通过十三行大量向海外输出，成为欧洲人心中中国漆艺的代表，被称为“广器”。

“广器”的主要品类有漆屏风、漆箱、金漆木雕、漆盒、女红桌、漆扇、餐桌以及定制的名片盒、信件盒等，采用黑底描金的设计，然后贴金或泥金，描绘出花草树石、亭台楼阁、人物典故、飞禽走兽等纹饰，以反映广州城内风光的题材为主要内容，如赛龙舟、广州十三行风景等，充分满足了西方人对中国的好奇心。还有的外商直接在欧洲做好家具木胎，然后千里迢迢运到广州进行髹漆彩绘后再运回，足见西方人对广州髹漆工艺的狂热。

Lacquer wares made in Guangdong province first appeared in the Warring States period, reached its peak in the Ming and Qing dynasties, and were represented by the golden lacquered wood-carving made in Chaozhou, lacquered furniture made in Canton and Foshan and lacquer wares made in Yangjiang. In the Qing dynasty, lacquer wares made in Guangdong province were exported overseas through the Thirteen Hongs, and became the representatives of Chinese lacquer wares in the European's mind, known as “Cantonese lacquer wares”.

The main categories of “Cantonese lacquer wares” are lacquer screens, lacquer cases, golden lacquer wooden sculptures, lacquer boxes, needlework tables, lacquer fans, dining tables, customized business card boxes, letter boxes, and so on, which were designed with black background with attached or painted and golden patterns of flowers, grass, trees, stones, pavilions, terraces, figures, stories, birds and beasts. The major themes are sceneries of Canton, such as dragon boat racing, and the

September / 2021.9

27

星期一

Monday

辛丑牛年 丁酉月

八月廿一

view of the trading region, satisfying Westerners' curiosity about China. There were also foreign traders who directly ordered wooden furniture in Europe, and then transport them to Canton to be lacquered and painted. That shows the Westerners were obsessed with the lacquer wares made in Canton.

19世纪通草水彩茶叶贸易图。（原件藏于广州十三行博物馆）

Tea trade, watercolour on pith paper, 19th century. (The original piece is collected in Guangzhou Thirteen Hongs Museum.)

September / 2021.9

28

星期二

Tuesday

辛丑牛年 丁酉月

八月廿二

广式家具：广州“西洋热”

Cantonese Furniture: “The Love for the Western Style” in Canton

广式家具指广州出产的木制家具，具有独特风格，是清式家具的典型款式之一，因而又被称为“广作”，与“苏作”“京作”“仙作”齐名。17 世纪末起，由于海禁的开放和一口通商的实施，大量外商及传教士来到广州，将西方文化元素也带至此地，掀起一股“西洋热”，对广州地区家具、建筑、绘画等都产生了影响。广式家具受西方巴洛克与洛可可式艺术风格的影响，追求用料充裕，采用大面积雕刻及镶嵌螺钿、玻璃油画等。构件一般不拼接，而是一木制成，为的是木性一致。装饰题材和纹饰也深受西方文化影响，出现束腰、兽脚、羊蹄腿、西番莲纹，并加重了对家具腿足的雕刻等，形成了广式家具用料粗大、体质厚重、雕刻繁缛的特点。

Cantonese furniture refers to the wooden furniture produced in Canton. It has a unique style and is one of the typical styles of the Qing furniture. Therefore, it is also called “Guangzuo” (works of Canton), as famous as “Suzuo” (works of Suzhou), “Jingzuo” (works of Beijing) and “Xianzuo” (the classical works). Since the end of the 17th century, due to the lifting of the ban on the maritime trade and the implementation of one-port trade, a large number of foreign traders and missionaries had been coming to Canton, bringing Western cultural elements here, arousing people's “following the western fashion” which had an impact on the Cantonese furniture, architectures, paintings, etc. Influenced by Western Baroque and Rococo art styles, the Cantonese furniture makers aimed to use materials in abundance, carve in large area, and inlay mother of pearl, oil painting on glass, and other objects. Components were generally not spliced,

September / 2021.9

29

星期三

Wednesday

辛丑牛年 丁酉月

八月廿三

but made of a whole piece of wood, in order to achieve texture consistency. The decorative themes and patterns were also deeply influenced by Western culture, and there appeared shapes like slim waists, animal foots, legs with sheep's hooves, and Western floral patterns. The artisans also emphasized the carving of the furniture's legs and feet, and the Cantonese furniture was characterized by its thick material, heavy weight and complicated carved patterns.

清代外销画：广州十三行的西式家具铺。（图片来源：*The China Trade——Export Pantings, Silver and Other Objects.*）

An export painting of the Qing dynasty: a Cantonese shop in the trading region that sold Western furniture. (The picture is taken from *The China Trade——Export Pantings, Silver and Other Objects,* by Carl L. Crossman.)

September / 2021.9

30

星期四

Thursday

辛丑牛年 丁酉月

八月廿四

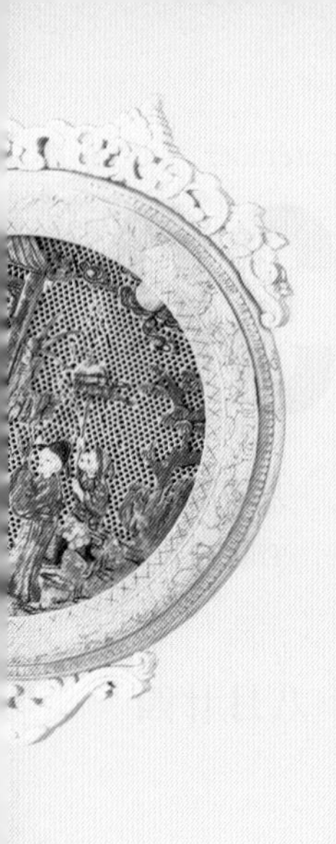

October

一	二	三	四	五	六	日
				1 国庆节	2 廿六	3 廿七
4 廿八	5 廿九	6 九月	7 初二	8 寒露	9 初四	10 初五
11 初六	12 初七	13 初八	14 重阳节	15 初十	16 十一	17 十二
18 十三	19 十四	20 十五	21 十六	22 十七	23 霜降	24 十九
25 二十	26 廿一	27 廿二	28 廿三	29 廿四	30 廿五	31 廿六

“广木作”：光耀紫禁城

Cantonese Furniture Made for the Court

广式家具因华丽繁复的风格，十分符合统治者追求豪华、繁复、绚丽的审美，而一度受到清朝宫廷、官绅和文人的追捧。据说光绪年间，广东名匠梁阜还曾被召前往北京，专门为皇帝大婚雕刻组装龙床。

自清代中期起，清皇室每年都从广州定做、采购大批家具，不仅如此，清宫造办处还专设“广木作”，挑选优秀的广州木工为皇室制作家具。“广木作”的工匠与“油木作”和“珐琅作”工匠合作，为清代宫殿、寺庙制作了屏风、佛龛、盒座、香几等大量器具，其中不乏“紫檀木西洋四方龛”“掐丝珐琅玻璃油画格子”等中西合璧的精品。

Because of its gorgeous and intricate style, Cantonese furniture was very suitable for the rulers' pursuit of luxury, complex and gorgeous aesthetics. Therefore, it was sought after by the Qing court, officials, gentry and scholars. It was said that during the Emperor Guangxu's reign, Liang Fu, a famous craftsman in Guangdong province, was once called to Beijing to carve and assemble a bed with dragon patterns for the emperor's wedding.

Since the middle of the Qing dynasty, the Qing royal family had ordered and purchased a large number of furniture from Canton every year. In addition, the royal workshop of the Qing court had also set up the “Guangmuzuo” (Cantonese furniture workshop), and selected excellent Cantonese carpenters to make furniture for the royal family. The craftsmen of “Guangmuzuo” cooperated with the craftsmen of “Youmuzuo” (lacquer ware workshop) and “Falangzuo” (enamel workshop) to produce a large number of furniture, including screens, niches, box seats and side tables for placing the incense burners for the palaces and temples in the Qing dynasty. There

October / 2021.10

1

星期五

Friday

辛丑牛年 丁酉月

八月廿五

国庆节

National Day

are many excellent works of Sino-Western styles, such as “square Western niche made of red sandalwood” “glass lattice with filigree enamel and oil painting”, etc.

清代黑漆描金庭院人物纹女红桌。女红桌主要由双门柜、针线盒、桌子三部分组成，通体黑漆描金装饰。柜内分大小不一的10个储物格，针线盒装饰与柜子一致，盒内 15 个隔断大小不一，内盛象牙制线轴、线梭、顶针等女红工具。19 世纪，附有象牙针线工具的女红桌是颇受欢迎的外销品。这张女红桌工艺制作精细，兼具女红桌与梳妆台功能，款式受西方文化影响，是中国外销家具中的精品。

Black lacquer gold-painted needlework table, Qing dynasty. The black lacquer needlework table, decorated with golden patterns, is mainly composed of three parts: the cabinet with two doors, the sewing box and the table. There are 10 storage compartments of different sizes in the cabinet. The decoration of the sewing box is in accordance with that of the cabinet. There are ivory thread shafts, thread shuttles, thimbles and other needlework tools in the box which is divided into 15 compartments of different sizes. In the 19th century, the needlework tables with ivory tools were popular export products. This needlework table is made with fine techniques, bearing the function of a needlework table and a dressing table. Its style is influenced by Western culture, and it is a fine piece among the Chinese furniture made for export.

October / 2021.10

2

星期六

Saturday

辛丑牛年 丁酉月

八月廿六

19世纪酸枝镂空雕竹节纹扶手靠背椅。

Chair with pierced patterns of bamboo, 19th century.

October / 2021.10

3

星期日

Sunday

辛丑牛年 丁酉月

八月廿七

镶嵌成精工

The Inlay Craft

清代十三行贸易的兴盛带动了瓷器制造、金银器嵌宝石等工艺技术的发展，为广式家具镶嵌工艺水平的提升创造了条件。广式家具的两大装饰技法，一是雕刻，一是镶嵌，行话称为“卖花”和“卖石”。镶嵌工艺大约在明代开始应用于广式家具中，到清代吸收了大量西方纹饰和技法而臻至成熟。清代广式家具的镶嵌艺术，深受欧洲金属嵌铜工艺、嵌胎珐琅工艺和蚀刻玻璃画工艺影响，发展出平嵌法、凸嵌法、框式镶嵌法和蚀刻玻璃画法工艺，把大理石（云石）、玉石、螺钿、珐琅、陶瓷、玻璃等镶嵌到家具中，将装饰性和功能性结合在一起，形成独树一帜的广式风格，风靡全国。

The prosperity of the Canton trade in the Qing dynasty led to the development of porcelain manufacture, gold and silver wares inlaid with gemstones and other crafts, creating conditions for the improvement of inlay technology of Cantonese furniture. The two major decorative techniques of Cantonese furniture were carving and inlaying, called “selling flowers” and “selling stones” in the jargon. The inlay technique was applied in the Cantonese furniture in the Ming dynasty, and it became mature in the Qing dynasty by absorbing a lot of Western patterns and techniques. The inlay art of Cantonese furniture in the Qing dynasty was deeply influenced by the European technologies of metalworks inlaid with copper, embossed enamel and etched glass painting. Hence, there appeared techniques such as flat-inlay, convex-inlay, frame-inlay and etched glass painting. Marble, jade, mother of pearl, enamel, porcelain and glass were inlaid into the furniture, to combine decoration and functionality, and form the unique Cantonese style which became popular all over the country.

October / 2021.10

4

星期一

Monday

辛丑牛年 丁酉月

八月廿八

清晚期摄影作品，展示了西关富商家中的广式家具。

A photograph taken in the late Qing dynasty that shows the Cantonese furniture of a rich family.

October / 2021.10

5

星期二

Tuesday

辛丑牛年 丁酉月

八月廿九

螺钿多姿彩

The Colourful Mother of Pearl

螺或贝类的壳体，因其内面能在光照下反射五彩光泽，被人们处理成片状，再加工成花鸟、树木、草叶及其他吉祥图案或形状，通过镶嵌或黏合方式施于器具表面，这些加工后的图案或形状称为螺钿。螺钿工艺据传起源于商周时期，最早用于漆器装饰，唐代则多施于铜镜背面，称为铜镜漆背螺钿。清代家具采用螺钿装饰的做法很普遍，从苏式到广式都有，一般多施于珍贵的硬木家具上，如紫檀家具、黄花梨家具、红木家具等。苏式家具的螺钿多与象牙、玉石、珊瑚等其他镶嵌物搭配使用，称为“百宝嵌”。而广式家具则特别喜欢使用螺钿工艺，椅凳、床榻、橱柜，不一而足，且多为满钿，使之色彩更显富丽堂皇和奇妙典雅。

The inner surface of the shell of a snail or shellfish, which can reflect colorful luster under the sunlight, are cut into pieces and then processed into shapes of flowers, birds, trees, grass, leaves and other auspicious patterns and applied to the surface of the objects by inlaying or attaching, This processed piece is called “Luodian” (the mother of pearl). The craft of inlaying is said to have originated in the Shang and Zhou dynasties. It was first used to decorate the lacquer wares. In the Tang dynasty, it was mostly applied to the back of the bronze mirror, which was called the bronze mirror with lacquered back and mother of pearl. In the Qing dynasty, the use of mother of pearl as decoration was very common, and could be seen in Suzhou style and Cantonese style furniture. It was generally applied to precious hardwood furniture, such as furniture made of red sandalwood, yellow-pear wood, redwood, etc. The mother of pearl in Suzhou style furniture is often used with other inlays such as ivory, jade, coral, etc., and this is called “Baibaoqian”

October / 2021.10

6

星期三

Wednesday

辛丑牛年 丁酉月

九月初一

(inlay of one hundred precious objects); while Cantonese furniture is especially inlaid solely with mother of pearl. Chairs, beds, cabinets, and many other objects are mostly inlaid with mother of pearl, so that the furniture's colour looks more magnificent, wonderful and elegant.

October / 2021.10

7

星期四

Thursday

辛丑牛年 丁酉月

九月初二

清末酸枝嵌螺钿框广绣花鸟纹插屏。屏心以米白缎为地，满工绣花鸟图案，梅花、菊花、牡丹竞相开放，鸟雀在枝头喧闹争鸣，鸭戏于水中，整幅画面生趣盎然，针艺高超。屏座以螺钿镶嵌缠枝花卉、福禄寿三星、梅花、蝴蝶等纹样，寓意吉祥。（原件藏于广州十三行博物馆）

Screen with Canton embroidery of birds and flowers, late Qing dynasty. A white satin is framed inside the screen. The satin is embroidered with patterns of birds, plum blossoms, chrysanthemums and peonies, which are blooming in competition to flaunt their beauty. Birds are noisy on the twigs; ducks are playing in the water. The whole picture is lively and the needlework technique is superb. The screen seat is inlaid with mother of pearl patterns of twigs, flowers, immortals of fortune, merit and longevity, plum blossoms, butterflies, etc., implying auspiciousness. (The original piece is collected in Guangzhou Thirteen Hongs Museum.)

广州木雕：神工精造成

Cantonese Wood-Carvings: Objects of the Fine Craft

广式家具和建筑讲究华丽堂皇，因而广州木雕也以精巧细腻著称。广州木雕多指硬木木雕，分建筑雕刻和家具雕刻两类，前者包括建筑的梁柱、屏风、门窗和栏杆，后者以红木家具和棒木箱为代表，此外还有宫灯及摆件。

广州木雕样式丰富，匠人所用的雕刻工具也十分繁多，主要分为毛坯刀和修光刀，还有圆凿刀、平刀和斜刀。一件正常大小的木雕，大约需要 35 件工具；假如是大型木雕，则需要 100 多件工具。在长期的发展过程中，广州木雕工匠还总结出了“精、绝、古、神、形、态”六字箴言。留存至今的陈家祠正是广州木雕的集大成者，其木雕种类丰富、精致多彩，充满浓浓的岭南文化韵味。

Cantonese furniture and architecture are exquisite and magnificent, while Cantonese wood-carvings are also famous for their delicacy. Cantonese wood-carvings mainly refer to the hardwood-carvings, which can be divided into architectural-carvings and furniture-carvings. The former include beams and columns, screens, doors, windows and railings of buildings, while the latter are represented by black wood furniture and wooden boxes, as well as palace lamps and ornaments.

Cantonese wood-carvings are diversified in terms of styles, and there are many carving tools used by the craftsmen: chisels and polishing knives, round chisels, flat knives and tilting knives. To make a normal size wood-carving, around 35 tools are needed; while to make a large wood-carving, more than 100 tools are needed. In the long-term development process, Cantonese carpenters also summed up “six essential rules” of the craft: to make it refined, unique, classical, spirited, finely-shaped and well-arranged. The Chen Clan Family that still remains today is a great evidence that keeps the Cantonese wood-carvings which are diversified in types, are exquisite, colorful, and full of Lingnan culture charm.

Cold Dew

寒露

October / 2021.10

8

星期五

Friday

辛丑牛年 戊戌月

九月初三

清末透雕酸枝框广绣松鹤延年图四屏屏风。屏风主体使用了酸枝木，中间镶嵌广绣松鹤延年图。绣品以黑缎为地，主要施金银色丝线，层次分明，针法多样且针路严谨。鹤的眼、嘴、脚尤其生动，活灵活现，体现了高超的刺绣技艺。在 18 世纪欧洲“中国风”盛行时期，上流社会热衷用中国的屏风、插屏等装饰家居。

Four-folded screen with Canton embroidery of cranes and pine trees, late Qing dynasty. The main body of the screen is made of black wood, and an embroidered picture of pines and cranes is inlaid in the middle. The embroidery is made of black satin. mainly adopts used gold and silver threads with various strict needling methods to create clear layers. The eyes, beaks and claws of cranes are particularly vivid and full of life, reflecting the embroiderers' superb skills. In the 18th century when “Chinoiserie” prevailed in Europe, the upper class people were keen to decorate their houses with Chinese ground-based screens and seated screens.

October / 2021.10

9

星期六

Saturday

辛丑牛年 戊戌月

九月初四

清末酸枝雕花红筋石面六角形高几。

End table with hexagonal stone panel and carved floral patterns, late Qing dynasty.

October / 2021.10

10

星期日

Sunday

辛丑牛年 戊戌月

九月初五

华贵酸枝街，盛名三友堂

The Redwood Workshops

早在宋代，广州玉带濠一带便是有名的娱乐消费场所，至清初，玉带濠濠畔街已是广州最旺的商店街，尤其以家具制作店居多。那时，紫檀、酸枝、花梨、坤甸等硬木从玉带濠水运至濠畔街，促使这里的酸枝木器作坊越办越多，达数十家，因而被称为“酸枝街”。

清末，广州地区最有名的木雕店是由广州许氏、佛山何氏和三水赵氏合营的“三友堂”。三友堂制作的木雕以雕工雄浑流畅、刀法洗练刚劲、格调高雅、装饰性强等特点而名扬省港澳。如今省港澳三地乃至东南亚留存的清末祠堂、寺庙及知名建筑的木雕，很多出自三友堂手笔。当时知名木雕店还有“伍常”和“万全”，也有很多木雕制品出口到海外。

Early in the Song dynasty, the area along Yudai Hao (Yudai River) in Canton was a famous place for entertainment and consumption. By the early Qing dynasty, the Riverside Street near the Yudai River had become the shopping centre in Canton, where there were many furniture shops. At that time, red sandalwood, redwood, scented rosewood, pontianak wood and other types of hardwood were transported to this street by the waterway of the Yudai River, so there appeared more and more workshops selling redwood. There were dozens of workshops in the street, so it was called the “Street of Redwood”.

At the end of the Qing dynasty, the most famous wood-carving workshop in Canton was the “Hall of Three Friends” jointly operated by the Xu family in Canton, the He family in Foshan and the Zhao family in Sanshui. The wood-carvings made by this workshop were famous in Hong Kong and Macao for the carpenters' powerful and smooth carving, neatly vigorous chiseling technique,

October / 2021.10

11

星期一

Monday

辛丑牛年 戊戌月

九月初六

the elegant style and strong decorativeness. Nowadays, the wood-carving objects in the temples and well-known buildings in the late Qing dynasty in these places and even in Southeast Asia were produced by this workshop. At that time, there were also other famous workshops such as "Wuchang" and "Wanquan"; many wood-carving products were exported overseas.

陈家祠中的《三顾茅庐》窗棂木雕画。（图片来源：FOTOE图片库）

A wood-carved window lattice: *Three Visits to the Cottage*, a scene from *The Romance of the Three Kingdoms*. (The picture is bought from the photo gallery FOTOE.)

October / 2021.10

12

星期二

Tuesday

辛丑牛年 戊戌月

九月初七

潮州金漆木雕：木作描金漆

Golden Wood-Carvings in Chaozhou

在“广作”中，潮州金漆木雕结合了漆器工艺和木雕工艺，显得十分特别。金漆木雕起源于唐代的潮汕地区，是一种建筑构件装饰艺术，工匠们选用樟木、银杏、冬青、杉木等为原料，在其上以沉雕、浮雕、圆雕、通雕、立体雕和锯通雕等技法进行雕刻，然后髹以生漆和金箔，或贴金而成，整个过程包括凿坯、细刻、磨光、揉漆、贴金等工序，十分复杂。

潮州金漆木雕的最大特点就是能在一块木头平面上雕出富有层次和立体感的图案，这要求工匠不仅具有精湛的雕工，还能够在脑海中想象出三维立体的图案造型。多层镂通的木雕鱼虾蟹篓是潮州金漆木雕中的精品，其代表作是著名艺人张鉴轩为北京人民大会堂广东厅设计制作的大型金漆木雕《花鸟》和《鱼虾》。

Among “the furniture made in Guangdong”, the wood-carvings with golden lacquer in Chaozhou combine lacquering and wood-carving techniques, and thus are very special. Originated in Chaoshan area in the Tang dynasty, wood-carvings with golden lacquer were a kind of decorative artworks used as architectural components. Carpenters choose camphor, ginkgo, evergreen, fir, and so on as raw materials, and carve on the wood with techniques such as intagliated carving, low-relief carving, full-relief carving, pierced carving, three-dimensional carving and saw-pierced carving, and then paint it with lacquer and golden foil or attached gold onto it. The whole procedure, highly complex, is divided into chiseling, carving the details, polishing, painting, pasting gold and other processes.

The most conspicuous feature of Chaozhou’s wood-carvings with golden lacquer is that the carpenters can create the patterns with multiple layers and a sense of realism on a piece of wood. That not only requires a carpenter to master

October / 2021.10

13

星期三

Wednesday

辛丑牛年 戊戌月

九月初八

fine carving techniques, but also to imagine the three-dimensional shapes in his mind. The wood-carving "a basket full of fish, shrimps and crabs", carved by the multi-layered piercing technique, is a fine piece of Chaozhou's wood-carving with golden lacquer. The designer of it is Zhang Jianxuan, a famous artisan who had designed the large-scale wood-carving works *flowers and birds* and *fish and shrimps* for the Guangdong Hall inside the Great Hall of the People in Beijing.

清代金漆木雕彩漆画菱形馔盒。

Rhombus lacquer box with golden painting, Qing dynasty.

October / 2021.10

14

星期四

Thursday

辛丑牛年 戊戌月

九月初九

重阳节

Double Ninth Festival

广州牙雕："关口象牙堆" Ⅰ

Cantonese Carved Ivory: the Ivory is Piled up in the Gateway. Ⅰ

广州人对于象牙制品的嗜好可以追溯到秦汉时期，当时人们认为象牙能够辟邪除秽，因而历代海上丝绸之路的舶来品中基本都有大量象牙，由巧手的工匠制作成算筹、牙雕等制品。唐代王建曾在诗里记录了广州象牙进口的盛况："戍头龙脑铺，关口象牙堆。"明代，在今解放中路和惠福路交界处形成了一条象牙商贸街。

广州牙雕兴盛于明代后期，在清代成为极具地方特色的民间工艺。清代广州一口通商后，来自于东南亚的象牙全部在广州交易，为牙雕工艺提供了充足的原料，广州牙雕由此获得长足发展，形成了自己的独特风格，工艺水平远超苏州、北京、扬州、杭州，成为全国之冠。

Cantonese people's preference for ivory products can be traced back to the Qin and Han dynasties, when people thought that ivory could ward off the evil spirits and eliminate filth. Therefore, among the imported products of the "Maritime Silk Route" in the past dynasties, there was a large number of ivory made into products such as arithmetic chips and ivory-carvings by skillful craftsmen. In the Tang dynasty, Wang Jian, wrote about the grand occasion of Canton's Ivory import in his poem: "there is a lot of borneol in the harbor, and the ivory is piled up in the gateway". In the Ming dynasty, an ivory trade street was formed at the junction of today's Jiefangzhong Road and Huifu Road.

Cantonese ivory-carvings flourished in the late Ming dynasty and became a folk craft with local characteristics in the Qing dynasty. After the one-port trade policy was implemented in Canton in the Qing dynasty, all the ivory from Southeast Asia was traded in Canton, providing sufficient raw materials for the art of ivory-

15

星期五

Friday

辛丑牛年 戊戌月

九月初十

carving. Therefore, Cantonese ivory-carving had made great progress and formed its own unique style. The technical level was far higher than that of Jiangsu, Beijing, Yangzhou and Hangzhou, and became the top all over the country.

19世纪后期到20世纪初期制作的酸枝木镂雕龙纹扶手椅。

Armchair with carved patterns of dragons, late 19th century to early 20th century.

October / 2021.10

16

星期六

Saturday

辛丑牛年 戊戌月

九月十一

晚清酸枝雕花高几。

End table with carved decorations, late Qing dynasty.

October / 2021.10

17

星期日

Sunday

辛丑牛年 戊戌月

九月十二

广州牙雕："关口象牙堆" Ⅱ

Cantonese Carved Ivory: the Ivory is Piled up in the Gateway. Ⅱ

广州牙雕又称"南派牙雕"，注重雕工，讲究牙料漂白和染色，以牙质莹润、精镂细刻见长，整体布局繁复热闹，不留空白。

广州牙雕的工艺有雕刻、编织和镶嵌三种。雕刻包括阴刻、隐起、起突、镂雕、染色，以镂雕最为典型，手艺精湛的工匠可以做到牙片薄如半透明的纸，雕镌细若游丝。编织指将经过特殊处理的象牙切割成丝，再编织成扇子等各种器物。镶嵌则是把象牙雕刻与紫檀、犀角、玳瑁、翠羽等巧妙地镶嵌于一器之上，使图案更富于层次。此外，广州牙雕还体现出鲜明的中西合璧风格，其图案纹饰除了花木、山石、龙舟、宝塔、蟹笼等岭南山水景物外，还有叶状纹、贝壳纹及写实折枝花卉等巴洛克纹饰。

Cantonese ivory-carving, also known as "ivory-carving of the southern school", emphasizing the processes of carving, bleaching and dyeing, is known for its soft texture, delicacy, complex and intricate composition without blank space.

In Cantonese ivory-carving, there are three kinds of techniques: carving, weaving and inlaying. The carving technique includes engraving, concealing, protruding, piercing and dyeing, among which, piercing is the most typical one. Skillful craftsmen can make an ivory piece as thin as translucent paper and a carved pattern as thin as silk. Weaving means to cut the specially processed ivory into threads, and then weave them into various objects such as fans. Inlaying means to inlay the carved ivory and other pieces, such as red sandalwood, rhinoceros horn, hawksbill shell, green feathers, into the objects so as to create the multi-layered patterns. Moreover, Cantonese ivory-carvings also embody a distinctive Sino-

October / 2021.10

18

星期一

Monday

辛丑牛年 戊戌月

九月十三

Western style. In addition to flowers, trees, rocks, dragon boats, pagodas, crab cages and other Lingnan landscapes, there are also Baroque patterns such as leaves, shells and flowers of realistic style.

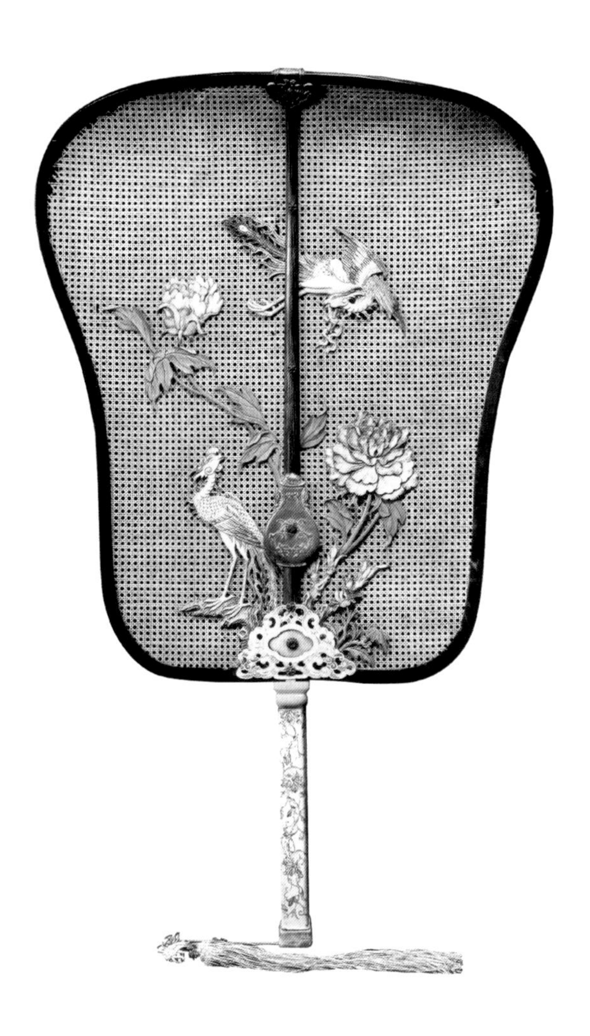

October / 2021.10

19

星期二

Tuesday

辛丑牛年 戊戌月

九月十四

牙丝官扇。清代广州的象牙艺师，利用象牙细致的纹理与不易碎的韧性原理，将用药水浸泡过的象牙劈成厚薄宽窄均匀的薄片，一直磨制到象牙呈现洁白的光泽为止，并用这项工艺制作成象牙扇。为增加扇面的立体感，扇面往往配有浮雕的吉祥花卉图案。（图片采自《中国扇具》）

Ivory fan made for the court. In the Qing dynasty, the Cantonese ivory artisans made use of the fine texture and toughness of the ivory, split the ivory which had been soaked in the liquid medicine into slices of even thickness and width, and polished them until the ivory slices showed a white luster. With these techniques, they weaved the slices into ivory fans. In order to increase the three-dimensional effect, the fans were often added with relief auspicious flower patterns. (The picture is taken from *The Chinese Fans*.)

广州牙雕：“关口象牙堆” Ⅲ

Cantonese Carved Ivory: the Ivory is Piled up in the Gateway. Ⅲ

广州牙雕大约在清代道光年间形成行会，分为贡行和洋行两种。贡行又称慎玉行，主要进行贡品生产和牙雕工匠管理，其产品须先由地方官或海关官员选购作为贡品，未中选者可自行销售；洋行又称为怀远堂，主要从事外贸生产，制作的摆件（象牙球、花舫、蟹笼、花塔等），日用品（折扇、台灯、烟嘴、笔筒、粉盒、图章、梳具、筷子、牙签、书签、纸刀等）和首饰（手镯、项链、耳环、戒指、别针等），极受西方消费者的欢迎。清末，在行会之外，又有一批自立门户的艺人或自学成才者成立大牙行，即“象牙会馆”。1958 年，“广州大新象牙厂”成立，生产的象牙球、象牙船、渔翁撒网代表了岭南牙雕的最高技艺。

Formed in the period of Emperor Daoguang in the Qing dynasty, the guilds of Cantonese ivory-carving could be divided into two types: the Tribute Hongs and the Western Hongs. The Tribute Hongs, also known as Shenyu Hongs, were mainly engaged in the production of the ivory paid as tribute and the management of ivory-carving craftsmen. The products must be selected by local officials or customs officials as tributes, while the products that were not selected could be sold by themselves. The Western Hongs, known as Huaiyuantang, were responsible for the products of foreign trade, among which, the ornaments (Ivory balls, ships, crab cages, floral towers, etc.), daily necessities (folding fans, table lamps, pipes, pen containers, powder boxes, stamps, combs, chopsticks, toothpicks, bookmarks, paper knives, etc.) and jewelry (bracelets, necklaces, earrings, rings, pins, etc.) were popular with western consumers. In the late Qing dynasty, in addition to these Hongs, there were also a group of self-supporting artists or self-taught talents who

October / 2021.10

20

星期三

Wednesday

辛丑牛年 戊戌月

九月十五

set up a big Hong, known as the Ivory Guild. In 1958, Guangzhou Daxin Ivory Factory was established and produced ivory balls, ivory boats and sculptures of fishermen scattering nets that represented the highest level of Lingnan ivory-carving.

19 世纪象牙雕花卉纹餐巾箍。（原件藏于广州十三行博物馆）

Ivory napkin buckle with pierced floral carving, 19th century. (The original piece is collected in Guangzhou Thirteen Hongs Museum.)

October / 2021.10

21

星期四

Thursday

辛丑牛年 戊戌月

九月十六

鬼工象牙球

Exquisite Ivory Balls

镂雕象牙多层套球是广州牙雕最经典的作品。牙雕工匠先从牙材中掏出一个球体，在球面往圆心处开出均匀分布的14个锥形孔，用直角钩刀在锥形孔中一层一层刻出活动自如的套球，再在每层球面上雕以各种精细复杂的纹样。由于制作繁复，对工艺要求极高，在宋代被称为“鬼工球”（当时可雕刻2层套球）。至清代乾隆年间，牙雕工匠已经能够雕出14层套球。19世纪出现了一件象牙球精品，直径约12厘米，共24层，表面以高浮雕刻九龙穿梭于祥云间，内部则为镂空锦地几何纹样。安放套球的支架如同烛台，同样用象牙雕成，底座为云龙纹，其上是镂雕八仙，中间为一颗雕有9层的云龙纹小套球，再上是山水人物，其工艺令人叹为观止。这件作品历经祖孙三代工匠、耗时百年才制作出，现藏于台北故宫博物院。

Multi-layered ivory ball is the most classical work of Cantonese ivory-carving. The artisan first scoops out a sphere from the ivory, cuts out 14 cone-shaped holes which will be evenly distributed on the surface and leading to the center of the sphere, then creates several movable layers through the cone-shaped holes with an orthogonal knife, and finally carves various delicate and complex patterns on each layer. Due to the complex technique and high requirements for the craft, it was called "ivory ball made by the spirit" in the Song dynasty (when the ball could only be carved with two layers inside). During the reign of Emperor Qianlong in the Qing dynasty, the artisans were able to carve 14 layers inside an ivory ball; in the 19th century, there appeared an exquisite ivory ball of 24 layers with a diameter of about 12 cm. The surface of the ball is carved in high-relief, with patterns of nine dragons shuttling among the auspicious clouds, while the interior of the ball is carved with

October / 2021.10

22

星期五

Friday

辛丑牛年 戊戌月

九月十七

pierced brocade of geometrical patterns. The bracket for holding the ball is like a candlestick, which is also carved out of ivory. The base is decorated with patterns of clouds and dragons, above which there are eight immortals. In the middle, there is a small ball of nine layers with patterns of clouds and dragons, and in the upper part there are landscapes and figures. The craftsmanship is amazing. It took three generations of craftsmen, namely about a hundred years, to finish this work. It is now collected in the Palace Museum of Taipei.

Frost's Descent

霜降

October / 2021.10

23

星期六

Saturday

辛丑牛年 戊戌月

九月十八

民国16层花卉纹象牙球连牙座摆件。16层象牙球外壁饰高浮雕牡丹花卉，镂雕14个圆孔，内层牙球可层层转动。柱身中部饰束腰柱体，底部分两层装饰，上下镂通雕喜上眉梢、祥云蟠龙纹，寓意吉祥。象牙球又称“鬼工球”，取“鬼斧神工”之意，是广州最具代表性的牙雕制品。

Ivory ball of sixteen layers with stand, period of the Republic of China. The outer layer of the 16 layered ivory ball is decorated with high-relief peony patterns, and there are 14 pierced round holes, in which all the inner layers can rotate. The middle part of the column stand is bound by a “waist-binding” band, and the base part is carved with two layers of decorations. The upper and lower parts are carved with magpies standing on the plum blossom twigs, and dragons playing among the clouds, which symbolizes auspiciousness. The ivory balls, also known as “the balls made by the spirits” meaning “such a craft can only be made by the spirits”, is one of the most representative Cantonese carved ivory artifacts.

October / 2021.10

24

星期日

Sunday

辛丑牛年 戊戌月

九月十九

清末象牙镂雕双夔龙耳柳亭人物纹瓶。口沿饰镂空网格花卉纹；顶部以镂空六角星纹铺底，浅浮雕柳亭人物纹，两侧饰通雕双夔龙耳；中部和底部由24片梯形牙片与牙条两两镶嵌而成，牙片均采用透雕拉丝锦地浅浮雕技法，内雕柳亭人物；足沿饰菊瓣纹。此瓶用料珍稀，雕工玲珑精巧，是典型的广作牙雕风格。

Ivory vase with two Chinese dragon-shaped handles and pierced patterns of figures and pavilions, Late Qing dynasty. The rim is decorated with pierced latticed patterns; the neck is decorated with hexagrams as ground patterns, low-relief figures, willows and pavilions; two handles with Chinese dragon heads are on both sides of the neck. The middle and lower parts are made up of 24 trapezoid ivory pieces, on which the brocade grounds with patterns of willows, pavilions and figures are carved with low-relief technique. The rim of the bottom is decorated by the chrysanthemum patterns. As a typical Cantonese ivory craft, the vase is made with precious material and the carving technique is delicate and refine.

广州象牙微雕：方寸见天地

Cantonese Micro-Carving Ivory

广州象牙微雕出现于20世纪初，是广州牙雕中出现较晚的品种，但因其对工匠的技艺要求极高，能于方寸中见天地，所以也是广州牙雕最具代表性的工艺品之一。

象牙微雕的创始人是冯公侠，1932年，他在一粒米粒大小的象牙上刻出孙中山先生遗嘱全文154字，被时人誉之“神眼”；他还在64开纸页大小的象牙片上刻出多达25000字的《共产党宣言》。1941年，冯公侠将《大学》《中庸》《论语》《孟子》四书微刻在象牙扇面上，全文共6万余字，分为288行，堪称鬼斧神工。其子冯少侠也是微雕大师，曾在象牙米上雕出形象各异、栩栩如生的十八罗汉像，令人叹为观止。

Cantonese micro-carving Ivory appeared at the beginning of the 20th century. Though it is a category that appeared comparatively late, it is also one of the most representative handicrafts of Cantonese ivory-carvings, marked by its high technical requirements for the craftsmen and its ability to show the intricate details within a tiny space.

The founder of micro-carving was Feng Gongxia. In 1932, he carved Dr. Sun Yat Sen's will about 154 words on an ivory piece as small as a grain, and he was known as "the divine eye" at that time. He also carved *The Communist Manifesto* about 25,000 words on an Ivory piece as large as a piece of A7 paper. In 1941, Feng Gongxia carved The Four Books, including *The Great Learning*, *The Doctrine of the Mean*, *The Analects of Confucius*, and *Mencius* on an ivory fan. The full text is more than 60,000 words, divided into 288 lines, and can be regarded as a masterpiece. His son, Feng Shaoxia, is also a master of micro-carving. It is amazing that he once carved 18 arhats with different images on an ivory piece as small as a grain.

October / 2021.10

25

星期一

Monday

辛丑牛年 戊戌月

九月二十

October / 2021.10

26

星期二

Tuesday

辛丑牛年 戊戌月

九月廿一

清代象牙高浮雕亚字形开光人物花卉纹名片盒。盒身一面高浮雕牡丹纹饰，居中倭角形开光内饰高浮雕戏曲人物故事，边饰随形雕绳索纹一周；另一面镂雕柳庭人物，居中椭圆形留白开光，四棱透雕西洋花卉。此盒材料珍贵、雕工精巧、纹饰繁缛，具有典型的中西合璧的外销风格。（原件藏于广州十三行博物馆）

Ivory business card box with high relief patterns of figures and flowers, Qing dynasty. One side of the box is decorated with high-relief peony patterns, and the open window with clipped corners is decorated with a high-relief story scene of opera characters, while the rim is decorated with rope patterns. The other side is carved with characters, willows, and pavilions; there is an oval open window in the middle, and the four edges are carved with Western flowers. This box is precious in material, exquisite in carving and intricate in decoration, marked by a typical Sino-Western style of export artifacts. (The original piece is collected in Guangzhou Thirteen Hongs Museum.)

象牙花舫：“天工”巧制造

Ivory Flower Boats: Made by the Artisans from the Heaven

清代《广州通志》曾载：“谚曰：苏州样，广州匠。香犀、象、蜃、玳瑁、竹、木、藤、锡诸器俱甲天下。”既肯定当时广州工艺水平高超，同时指出广州工艺博采众长、兼容并蓄的特点。广州牙雕三绝之一的象牙花舫最初雕刻工艺正是吸收了苏州地区的核舟雕刻技法。

乾隆三年（1738 年），供职于清廷造办处的广州牙雕名师黄振效参考苏州核舟雕刻了一座高 1.7 厘米，长 5.2 厘米，宽 1.5 厘米的象牙镂雕御船。船身满施镂空雕刻，船首雕牌坊，牌坊后刻有带着 9 扇可以开合窗户的中空篷舱。此外，还雕了各具情态的 3 个坐船人和 7 个船工，并在船底置能左右活动的船舵，十分精巧。这艘船很可能是最早的象牙花舫，现藏于故宫博物院。

The General Comprehension of Canton written in the Qing dynasty once mentioned: “there is a proverb saying: the best design is from Suzhou, while the best craftsmen are from Canton. The Cantonese rhino horns, ivory, shell, hawksbill shells, bamboo, wood, rattan and tin are all the best in the world”. It points out that the Cantonese artisans were extremely skillful, and at the same time, they learned from others and are good at integrating the new elements into the original ones. The carving technique of the ivory flower boats, one of the three unique ivory-carving types in Canton, first absorbed the technique of making pit-carving boats in Suzhou area.

In the third year of Emperor Qianlong's reign (1738), Huang Zhenxiao, a famous Cantonese Ivory carver who worked in the royal workshop of the Qing court, learning and using the technique of pit-carving boats in Suzhou, carved an ivory imperial ship with a height of 1.7cm, a length of 5.2cm and a width of 1.5cm. The ship

October / 2021.10

27

星期三

Wednesday

辛丑牛年 戊戌月

九月廿二

is full of pierced carvings. The bow of the ship is carved with a memorial archway. Behind the archway, there is a hollow cabin with 9 windows that can be opened and closed. In addition, three passengers and seven boatmen with different facial expressions are carved, and a rudder that can be moved left and right is placed at the bottom of the boat. This extremely delicate ship was probably the earliest ivory boat, and is now collected in the Palace Museum.

October / 2021.10

28

星期四

Thursday

辛丑牛年 戊戌月

九月廿三

清代象牙透雕庭院人物天使纹盒。主体由四面透雕牙板相接而成，每面牙板中央均饰如意形开光，内饰柳亭人物纹，顶部为透雕如意形装饰；顶部中央为两个圆孔作提手，圆孔两侧饰扎发髻的中国风小天使和花卉纹。此盒形制特别，透雕工艺精细。小天使的题材不仅体现了广东民间艺人取材广泛、不受约束的特点，也是中西文化交流的见证。（原件藏于广州十三行博物馆）

Ivory box with pierced patterns of gardens, figures and angels, Qing dynasty. Four ivory plates with pierced-carving patterns are jointed together to form the box. In the center of each plate, there is a Ruyi-shaped open window decorated with characters, willows and pavilions. On the top of the box there is also a pierced Ruyi-shaped decoration, with two round holes as handle, beside which there are two Chinese angels with hair buns and floral patterns. This box is made in a special shape, with fine pierced-carving craftsmanship. The addition of little angels not only shows that the Guangdong folk artists' extensive subjects and unrestricted use of materials, but also witnesses the Sino-Western cultural exchange. (The original piece is collected in Guangzhou Thirteen Hongs Museum.)

广州玉雕：巧思在其中 Ⅰ

Cantonese Jade Carving: Marvellous Design Ⅰ

玉雕、牙雕和木雕是广雕中最负盛名的三种。其中玉雕是中国最古老的雕刻品种之一，早在距今六七千年的河姆渡文化时期就已经出现玉雕制品。广州玉雕在西汉时已有较高工艺水平，唐代中后期进入兴盛期。广州本地并不产玉，玉雕原料多为来自于粤西信宜的“南方玉”。也许是因为玉料不易得的缘故，广州工匠在雕玉时构思精巧、不拘一格，在宋代便出现了“七巧色玉”的雕刻技巧，后又在此基础上独创了“留色”技巧，保留下玉料的原色，雕刻出的玉器既精致又生动，充分展现了工匠的巧思。

Jade carving, ivory carving and wood carving are the three most famous carvings in Canton. Among them, jade-carving is one of the oldest category of Chinese carving, and the jade-carving products appeared early in the Hemudu cultural period of 6,000 years ago. Cantonese jade carving achieved a high technical level in the Western Han dynasty, and became popular in the middle and late Tang dynasty. The jade is not produced locally in Canton. Most of the jade-carving materials, namely “the southern jade”, are from Xinyi in the western part of Guangdong province. Perhaps because it is not easy to obtain jade, the Cantonese craftsmen are ingenious and unconventional in terms of design. In the Song dynasty, the carving technique of “jade of seven colours” appeared, and then on that basis, those craftsmen they created the technique called “preserving the colours” to retain the original colours of the jade. The carved jade objects are exquisite and vivid, showing the craftsmen’s ingenuity.

October / 2021.10

29

星期五

Friday

辛丑牛年 戊戌月

九月廿四

October / 2021.10

30

星期六

Saturday

辛丑牛年 戊戌月

九月廿五

19世纪象牙雕竹报平安纹名片盒。

Ivory business card box with auspicious patterns of bamboo, 19th century.

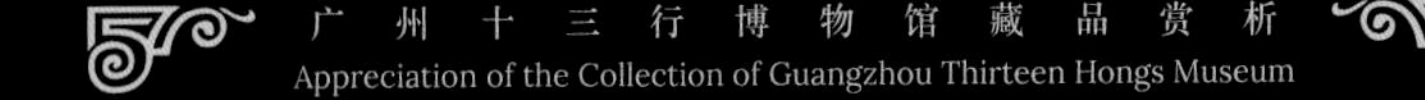

October / 2021.10

31

星期日

Sunday

辛丑牛年 戊戌月

九月廿六

清代象牙镂雕人物纹画框。画框通体镂雕，以仙鹤、老鹰、宝相花、卷叶纹为饰，图案两边对称，特别之处是雕刻了一对小天使，呈“八”字围绕在花卉两侧，做飞翔姿势；画框两旁则雕刻了中国传统柳亭人物纹；画框中间镶嵌一象牙板做画面，画面内容为彩绘人物。在中国风盛行的年代，欧洲人对这种既有中国传统寓意图案，又有西方宗教题材的工艺品情有独钟。

Ivory picture frame with pierced figural patterns, Qing dynasty. The whole picture frame is carved with pierced patterns such as cranes, eagles, precious flowers and rolling leaves. The patterns on both sides are symmetrical. In particular, on both sides of the central flower, a pair of little angels, whose positions represent as the Chinese character "eight"(八), are carved in the gestures of flying. The two sides of the picture frame are carved with Chinese traditional willows, pavilions and figures. An ivory board is inlaid in the middle of the picture frame as a painting, in which there are polychrome figures. In the era when Chinoiserie prevailed, Europeans had a special interest in this kind of handicrafts with both Chinese traditional allegorical patterns and western religious themes.

November

一	二	三	四	五	六	日
1 廿七	2 廿八	3 廿九	4 三十	5 寒衣节	6 初二	7 立冬
8 初四	9 初五	10 初六	11 初七	12 初八	13 初九	14 初十
15 十一	16 十二	17 十三	18 十四	19 下元节	20 十六	21 十七
22 小雪	23 十九	24 二十	25 廿一	26 廿二	27 廿三	28 廿四
29 廿五	30 廿六					

广州玉雕：巧思在其中 Ⅱ

Cantonese Jade-Carving: Marvellous Design Ⅱ

清初废除“匠籍”制度，兼之广州成为清朝唯一的对外贸易大港，来自全国的很多能工巧匠都汇聚于广州，推动了广州手工艺技术的飞速发展。广州玉雕汲取了各地玉雕技术，并横向与牙雕、木雕融会贯通，形成了独具一格的岭南玉雕风格。

广州玉雕分为首饰和摆件两大类，摆件中尤以玉球、花舫为特色。玉球类似象牙球，工匠们克服玉石硬度高、难加工的困难，在玉料上通雕出多达20层的玉套球，并雕刻山水、人物、虫鱼、花鸟作为装饰，薄厚均匀，转动自如。广州玉雕发展到20世纪30年代趋于鼎盛，玉雕业户达4000多家，长寿路、带河路玉器圩，成为我国南方最大的珠宝玉器市场；20世纪60年代末成立的南方玉雕厂，成为传承岭南玉雕工艺的主要基地和华南规模最大的玉器厂家。

In the early Qing dynasty, the “craftsmen register” system was abolished, and Canton became the only one major foreign trade port. Many skillful craftsmen from all over the country gathered in Canton, promoting the rapid development of Cantonese handicraft techniques. Cantonese jade-carving absorbed techniques from all over the country, and formed a unique Lingnan jade-carving style by integrating ivory and wood-carving techniques.

Cantonese jade-carving can be divided into two categories: jewelry and ornaments. The ornaments are especially represented by jade balls and flower boats. Jade balls are similar to Ivory balls. Artisans have to overcome certain problems such as the hardness of the jade and difficulty in processing. They carve up to 20 layers inside a jade ball, with landscapes, flowers, figures, insects, fish, flowers and

November / 2021.11

1

星期一

Monday

辛丑牛年 戊戌月

九月廿七

birds as decorations. Each layer is thin and even and can rotate freely. Cantonese jade-carving reached its peak in the 1930s, with more than 4,000 jade-carving workshops. Jade markets around Changshou Road and Daihe Road became the largest jewelry and jade market in South China; the Jade-carving factory in the south was established in the late 1960s, and became the main base for inheriting Lingnan jade-carving technique and the largest jade manufacturer in South China.

广州玉雕：岫玉云龙球。（图片来源：FOTOE 图片库）

Cantonese jade-carving: a jade ball with patterns of clouds and dragons. (The picture is bought from the photo gallery FOTOE.)

November / 2021.11

2

星期二

Tuesday

辛丑牛年 戊戌月

九月廿八

榄雕：核上乾坤大

Olive Pit-Carving: Intricate Patterns on the Pit

榄雕是果核雕的一种，特指以乌榄核雕成的工艺品，出现的时间不晚于明代，据载始创于增城新塘。新塘盛产乌榄，其榄核个头很大，质脆且具有超过象牙的硬度，细工雕刻不会留下刀屑，很适合作为榄雕材料。增城的榄雕水平很高，到了清代已成为地方官员每年进献宫廷的贡品。

榄雕的第一步是选材，根据榄核的大小、肥瘦、曲直、异变等外形进行构思，确定下雕刻造型；第二步是锉坯，用锉将原料锉出大致的形状，以便进行下一步；第三步是镂空成型，这是整个工序中最核心也最为耗时耗力的一步，业内有口诀云“下刀如鬼神，定夺死与生，刀尖有力稳，雕出精良品”；然后，再经雕花明线、镶嵌成件、打磨上蜡、上油、装木座，才能最终完成。

Olive pit-carving is a kind of carving on fruit pit, especially referring to the handicraft made of black olive pits, which appeared no later than the Ming dynasty. It was said that it originated in Xintang town, Zengcheng. Black olives abound in Xintang. The olive pit in Xintang is very big, brittle and harder than ivory. If an olive pit is finely carved, there will not be any scrap, so it is a suitable material for pit-carving. In Zengcheng, the olive pit-carving had reached very high level, and by the Qing dynasty, it had become an annual tribute paid to the court by local officials.

The first step of olive pit-carving is to select the material, design the composition according to the size, plumpness, straightness and shapes of the olive pit, and arrange the layout. The second step is to carve the embryo, which means to carve the raw material into the approximate shape for the next step. The third step, to mould the shape and hollow it out, is the core and the most time-consuming step

November / 2021.11

3

星期三

Wednesday

辛丑牛年 戊戌月

九月廿九

in the whole process, so there is a saying in the industry: "to carve as if you are a ghost and god, to determine death and life, to be powerful and stable when using the blade, and to create excellent products". Finally, it will not be finished until it is carved with patterns, inlaid with objects, polished, waxed, painted and installed in a wooden base.

广州榄雕：渔翁撒网。（图片来源：FOTOE 图片库）

Cantonese olive pit-carving: a fisherman scattering the fishnet. (The picture is bought from the photo gallery FOTOE.)

November / 2021.11

4

星期四

Thursday

辛丑牛年 戊戌月

九月三十

菊生湛盛名

Zhan Jusheng, a Famous Artisan

据说，历史上最出名的榄雕工匠是清代新塘人湛菊生。湛菊生原本是读书人，因屡试不第，后专心于榄雕创作，技艺驰名省港澳，并远扬南洋等地。清代咸丰年间，54 岁的湛菊生雕刻出了著名的“东坡夜游赤壁”花船，再现了苏东坡、黄庭坚和佛印三人并船家共游赤壁的场景。板寸见方的船底上，还有湛菊生刻的《前赤壁赋》537 字，衣针大小的橹杆上刻有“咸丰甲寅时年五十三谷生作”的落款，刀工之精令人惊叹。

现代榄雕作品继承传统，运用了镂雕、浮雕、圆雕、微雕等手法创作，有 50 多种品类，如多层花舫、云龙花瓶、通雕蟹笼、吊链宫灯、花塔、古鼎、国际象棋等。

It is said that the most famous olive pit-carving craftsman in history was Zhan Jusheng, a native of Xintang in the Qing dynasty. Zhan Jusheng was originally a student. Because he repeatedly failed in the imperial exams, he concentrated on olive pit-carving. His skills were well-known in Guangdong province, Hong Kong, Macao, and Southeast Asia. During the Emperor Xianfeng's reign in the Qing dynasty, Zhan Jusheng, aged 54, carved the flower boat, a famous work called "Dongpo's night tour to the red cliff", reproducing the scene in which Su Dongpo, Huang Tingjian and Foyin travelled together to the red cliff. On the bottom of the little boat, there are 537 characters from *The Ode to the former Red Cliff* carved by Zhan Jusheng, and the Tuscribe, which says "a work done at the age of fifty-three in the year of Jiayin (1854), during the reign of Emperor Xianfeng", carved on the pole as thin as a needle. The craftsmanship is so excellent that it astonishes people.

Modern olive pit-carving has inherited the tradition, applying means of

November / 2021.11

5

星期五

Friday

辛丑牛年 戊戌月

十月初一

pierced-carving, low-relief, round-carving and micro-carving. There are more than 50 categories, such as multi-layered flower boats, vases with patterns of clouds and dragons, pierced crab cages, hanging palace lamps, floral towers, ancient goblets, chess sets, etc.

清代银錾刻松竹梅人物纹瓜形双耳罐（铭文：和昌 HC）。

Silver pumpkin-shaped pot with two handles and patterns of pine trees, bamboo, plum blossoms and figures, Qing dynasty (inscription: He Chang HC).

November / 2021.11

6

星期六

Saturday

辛丑牛年 戊戌月

十月初二

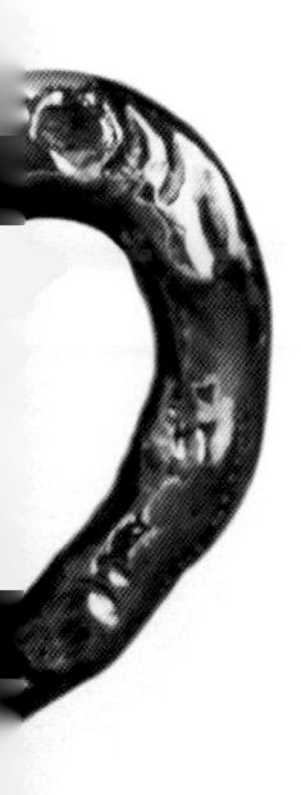

清代银“陈张李向巡查缉捕”三桅船模(连座)。船模重1724克,通体银质,使用锤揲錾刻技法。船上各部件一应俱全。船舵舵叶上开菱形小孔,转舵时能减少水的阻力;船帆是传统的平衡四角帆,可以适应风向转动。甲板上共有5门大炮,还有10名官府人员。船尾竖立四面船旗,分别錾刻“陈”“张”“李”“向”,船尾錾“巡查缉捕”。此船模工艺精湛,反映了清代缉捕船的形制,对于古代船舶的研究和外销银器的研究具有比较重要的参考价值。

立冬

Beginning of Winter

November / 2021.11

7

星期日

Sunday

辛丑牛年 己亥月

十月初三

Three-masted silver model ship (with a stand), Qing dynasty. The model ship weighs 1724 grams and is made of silver. The artisans used hammering and carving techniques to make all the parts of the ship. A rhombus hole is opened on the rudder blade, which can reduce the resistance of water when the rudder turns; the sail is a traditional balanced four-angled sail that can adapt to the wind direction. There are 5 cannons on the splint and 10 officials on board. At the stern, four flags respectively carved with the characters "Chen", "Zhang", "Li" and "Xiang" are erected, and the stern is carved with the characters "Patrol and Arrest". The craftsmanship of this model ship is refined. The model ship shows the shape of the patrol ships in the Qing dynasty and has an important referential value for the study of ancient ships and export silver wares.

中国外销银：风靡西洋地 Ⅰ

Chinese Export Silver Wares: Popular with the West Ⅰ

在十三行的外销品名单里，瓷器、丝绸、茶叶最为常见，而以银器为代表的金属工艺品则几乎不见记载。这并非是因为中国的金属工艺不好，恰恰相反，当时广州生产的银器是国际知名的高档广货，只是没有大批量生产，所以很少留下记录，在国际市场上，它被称为“中国外销银”。

清代，以外销银器为主的银器市场位于十三行商馆区的同文街、靖远街及河南花地一带，那时著名的银器铺有宝盈、“Tuhopp”、锦成、林盛、新时、侯昌等。欧洲在工业革命之后，银器使用量迅猛上涨，还因此出现了高额的金银制品税。而当时的中国不缺白银，人力成本又低，所以广东银器很快便以工艺精湛、价格低廉、质量上乘的优点赢得了西方人的喜欢。

Porcelain, silk and tea were the most common export products in the Canton trade, while the metal crafts represented by silver wares were rarely mentioned. This is not because China's metal crafts were not good. On the contrary, the silver wares produced in Canton at that time were high-end goods famous all over the world, but they were not made in mass production, so few records mentioned about them. In the international market, they were called “Chinese export silver wares”.

In the Qing dynasty, the silver market that mainly sold export silver products was located in Tongwen Street, and Jingyuan Street in the trading region as well as Huadi area on the southern bank of the Pearl River. At that time, the famous silver shops were Baoying, Tuhopp, Jincheng, Linsheng, Xinshi, Houchang, etc. After the industrial revolution in Europe, the use of silver wares increased rapidly, resulting in a high taxation on gold and silver products. At that time, China was not short of silver, and the labor cost was low, so silver wares made in Guangdong province

November / 2021.11

8

星期一

Monday

辛丑牛年 己亥月

十月初四

quickly won over the Westerners with the advantages of exquisite craftsmanship, low prices, high quality and high yield.

清代银镂空刻开光凤穿牡丹纹“福”字首饰盒。通体采用镂空锤揲錾刻技法，以不规则斜面为开口；盒盖、盒腹、盒背沿饰连珠纹，主体图案为单面镂雕凤穿牡丹；盒一侧系银链，方便携带；盒盖内侧錾“福”款。首饰盒雕工精美，造型中西结合，款式高雅大气，多放贵妇小姐的饰物。（原件藏于广州十三行博物馆）

Silver jewelry case with pierced patterns of phoenix, peonies and Chinese characters of "Fu", Qing dynasty. The artisans adopted the hammering and pierced-carving techniques to make the case, with the irregular tilted surface as the opening. The cover, the main part and the back of the case are decorated with beaded patterns, and the main pattern is a phoenix with peonies made by single-sided pierced carving; a silver chain is tied to one side of the case, which makes the case convenient to carry; the inside of the cover is carved with the Chinese character "Fu". The jewelry case, exquisitely carved, is made in a Sino-Western shape, with an elegant and wonderful style. The cases of this kind are mostly the ladies' ornaments. (The original piece is collected in Guangzhou Thirteen Hongs Museum.)

November / 2021.11

9

星期二

Tuesday

辛丑牛年 己亥月

十月初五

中国外销银：风靡西洋地 Ⅱ

Chinese Export Silver Wares: Popular with the West Ⅱ

目前已知最早的中国外销银器是一件制作于1685年的银盒，现藏于英国维多利亚和阿尔伯特博物馆。中国银器真正成为外销商品则从18世纪开始，可分为四个时期：早期海贸（1785年以前）、海贸时期（1785–1840年）、晚期海贸（1841–1885年）、海贸之后（1885年以后），每个时期各有特点。

早期海贸的外销银器存世很少，尚未形成风格特征；进入海贸时期，外销银器为满足西方市场需求，仿造西方银器形制纹样，以西洋风格为主，兼有少量中国风格的银器。1841年之后，外销银器的中国元素增多，广东银匠将中国传统文化中的龙凤、戏曲故事、人物风景、花鸟虫鱼等题材用于银器纹饰，浓郁的中国风情深受西方人青睐。

At present, the earliest known Chinese export silver ware is a silver box made in 1685, which is now collected in Victoria and Albert Museum in England. Since the 18th century, China's silver wares had really become export commodities. There were four periods with respective characteristics: early maritime trade period (before 1785), maritime trade period (1785-1840), late maritime trade period (1841-1885), and post maritime trade period (after 1885).

In the early stage of maritime trade, there were few export silver wares, which had not formed a style or feature; in the maritime trade period, in order to meet the needs of the western market, the artisans copied the shapes and patterns of western silver wares, the export silver wares were mostly made in western style, while a small number of them were made in Chinese style. After 1841, the Chinese elements were adopted to decorate the export silver wares. Guangdong silversmiths carved

November / 2021.11

10

星期三

Wednesday

辛丑牛年 己亥月

十月初六

dragons and phoenixes, opera stories, characters, landscapes, flowers, birds, insects, fish and other patterns from traditional Chinese culture as decorations on the silver wares. Such a rich Chinese style was favored by the westerners.

清代银錾刻人物纹瓜形奶壶。（原件藏于广州十三行博物馆）

Silver milk jug with figural patterns, Qing dynasty. (The original piece is collected in Guangzhou Thirteen Hongs Museum.)

November / 2021.11

11

星期四

Thursday

辛丑牛年 己亥月

十月初七

中国外销银：风靡西洋地 Ⅲ

Chinese Export Silver Wares: Popular with the West Ⅲ

外销银器的制作工艺包括范模浇铸、锤揲、錾刻、贴焊、镂空、累丝、鎏金、银胎珐琅、模压、镶嵌等，这样制作出来的银器器物造型纹饰极具立体感和真实感。当时西方银器的底部有代表银器成色、产地、年份和制作人的底款，广东银匠不懂其含义，也照葫芦画瓢随意标上，这被称为“伪款”。1865年之后，伪款渐趋消失，取而代之的是中国工匠或银楼的名字字母缩写和“90”“足纹”等字样。

广州作为外销银器最早诞生的地方，在19世纪中叶已根据客户群体的不同，将金银器制品细分为“唐行”和“洋货行”。唐行主营首饰、中餐具等中国传统银器；洋货行主营各种西式器具，如怀表附件、大型奖杯、西洋茶具和餐具等。

The manufacturing process of export silver wares includes molding, hammering, chiseling, fillet-welding, piercing, adding filigree, gilding, enameling, pressing, inlaying, etc. The patterns of the silver wares made in this way are three-dimensional and have a sense of realism. At that time, there were inscriptions representing the silver wares' precentage of silver, origins, years of production and producers at the bottom of the western silver wares. Silversmiths in Guangdong didn't understand the meaning of the inscriptions, but also imitated them and added the inscriptions, which were called “fake inscriptions”. After 1865, the fake inscriptions gradually disappeared, and were replaced by the abbreviations of Chinese craftsmen or silver workshops such as “90” and “Zuwen” (foot pattern).

By the middle of the 19th century, the gold and silver products had been divided into “Tang Hongs” and “Hongs of Foreign Goods” according to different consumer groups in Canton, the earliest birth place of export silver wares. Tang Hongs

November / 2021.11

12

星期五

Friday

辛丑牛年 己亥月

十月初八

were mainly engaged in selling jewelry, Chinese table wares and other traditional Chinese silver wares, while the Hongs of Foreign Goods were mainly engaged in selling various Western-style objects, such as pocket watches, accessories, large trophies, western tea sets, table wares, etc.

民国银累丝烧珐琅蝙蝠纹六角形果盒。

Silver filigree hexagonal box for preserving fruits with enamel patterns of bats, period of the Republic of China.

November / 2021.11

13

星期六

Saturday

辛丑牛年 己亥月

十月初九

November / 2021.11

14

星期日

Sunday

辛丑牛年 己亥月

十月初十

民国银錾刻竹纹镂空套玻璃酒瓶。

Glass wine bottle mounted by silver with pierced patterns of bamboo, period of the Republic of China.

外销扇：灿然有华风 Ⅰ

Export Fans: the Brilliant Chinese Style Ⅰ

在 17 至 18 世纪的欧洲上流社会，所谓的中国风尚渗透了日常生活的方方面面——花园要有中式亭台，建筑要有中国元素，穿中国丝绸，用中国瓷器喝中国茶叶，甚至连扇子也以用中国扇为荣。

欧洲人之所以青睐中国外销扇是有其独特的文化背景的。中世纪以来，扇子在欧洲被女性视为必不可少的服装点缀品，是身份、地位和礼仪的象征，因而价格昂贵。16 世纪，随着东西方航路的开通，制作精美、富有中国趣味而又价格相对低廉的中国外销扇进入欧洲，自然引发了追捧热潮。据记载，清光绪六年（1880 年）中国出口货品清单中，各种类扇子的数量达到 6,287,989 把，共值银 38,881 两。

In the upper class of Europe in the 17th and 18th century, the so-called Chinese fashion Permeated in all aspects of people's daily life: in the gardens there should be Chinese pavilions; in terms of architecture, there should be Chinese elements; when it came to clothing, Chinese silk was indispensable; when they drank, they used Chinese porcelain, and if they used fans, they would use Chinese fans.

There was a unique cultural background for the Europeans' preference for the Chinese export fans. Since the Middle Ages, fans had been regarded as the essential female decorations in Europe. They were symbols of identity, status and etiquette, so they were expensive. In the 16th century, with the opening of the routes connecting the East with the West, the exquisite export fans with Chinese style were sold to Europe at relatively low prices, and consequently were ardently pursued. According to the records, an inventory of Chinese export goods shows that in the sixth year of

November / 2021.11

15

星期一

Monday

辛丑牛年 己亥月

十月十一

Emperor Guangxu's reign(1880), the number of export fans of various types was up to 6,287,989, with a total value of 38,881 tael of silver.

清代黑漆描金骨纸本彩绘人物图折扇。21 支黑漆描金扇骨，两大骨嵌螺钿及描金花卉纹。扇面为双面纸本工艺，一面绘庭院山水人物图，另一面为素地纸本，中央绘仕女花卉图，风格清新雅致。这把扇以画工精细的冷色调画面，配上金光灿烂的扇骨，冷暖简繁互为对应，多种工艺相结合，富有装饰美。（原件藏于广州十三行博物馆）

Black lacquer folding fan with gold-painted polychrome figures on paper, Qing dynasty. The fan consists of 21 black lacquer sticks with gold-painted patterns, and two larger sticks are in laid with mother of pearl and painted with golden floral patterns. The fan is made with the double-sided paper-based technique. Figures in the landscape courtyard are painted on one side, and ladies and flowers are painted in the middle of the other side, which is based on paper. The style is fresh and elegant. The delicate cool tone of the painting on the fan and the sticks with brilliant golden patterns form a contrast between the cool and warm, and a comparison between the simple and complex. Many processing techniques are integrated to create a decorative beauty. (The original piece is collected in Guangzhou Thirteen Hongs Museum.)

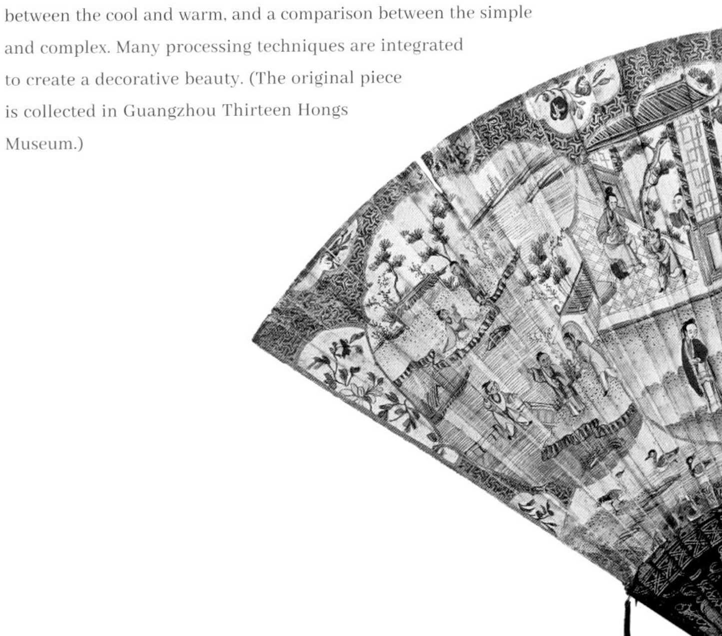

November / 2021.11

16

星期二

Tuesday

辛丑牛年 己亥月

十月十二

外销扇：灿然有华风 Ⅱ

Export Fans: the Brilliant Chinese Style Ⅱ

1699 年，英国东印度公司首次在广州订购了 8 万把外销扇，因工艺精美，材质名贵，极具东方趣味，很快销售一空，不仅在欧洲引发了一股中国风潮，也刺激了广州扇业的发展，推动广州制扇作坊数量迅速增长，时人称“城厢扇馆遍东西，花样之多折叠齐”。今天大新路、状元坊、德星路、长寿路一带是当时制扇作坊集中的地方，而且分工十分细致，比如“扇仔行”专做扇坯，“贡牙行”做扇骨拉花，“牛骨行”刨骨料，“雅风行”裱扇面，“仁风行”则做最后的组装工作，形成了一条庞大的产业链。

In 1699, the English East India Company ordered 80,000 export fans in Canton for the first time. Because of the exquisite craftsmanship, precious materials and oriental style, the fans sold out quickly, that not only led to a Chinese fashion in Europe, but also stimulated the development of the fan industry in Canton and promoted the rapid growth of fan-making industry in Canton. There was a saying describing the situation: “the fan-making workshops were all over in the city, there are so many different styles, and the fans can be folded neatly”. Today, Daxin Road, Zhuangyuanfang, Dexing Road and Changshou Road were the places where fan-making workshops were located at that time, and the fan-making process were meticulously divided into different categories. For example, “Shanzai Hong” specialized in making the prototypes, “Gongya Hong” carved patterns on the fan sticks, “Niugu Hong” dealt with the bone materials, “Yafeng Hong” mounted the fan surface, and “Renfeng Hong” finished the final assembly work. Thus a huge industrial chain was formed.

November / 2021.11

17

星期三

Wednesday

辛丑牛年 己亥月

十月十三

19世纪象牙雕扇骨花卉纹折扇。（原件藏于广州十三行博物馆）

Ivory folding fan with floral patterns on Canton embroidery, 19th century. (The original piece is collected in Guangzhou Thirteen Hongs Museum.)

November / 2021.11

18

星期四

Thursday

辛丑牛年 己亥月

十月十四

外销扇：灿然有华风 Ⅲ

Export Fans: the Brilliant Chinese Style Ⅲ

这些产自广州的外销扇与中国传统扇子有很大区别，它们造型丰富，色彩艳丽，工艺奇巧，材质多样，华美无比。

中国外销扇大致分为折扇和团扇两类，但其下又可根据造型细分为不同种类，比如折扇分有扇面的折扇和纯扇骨组成的折扇。有扇面的又分为对称和不对称、普通型和多层型，而纯扇骨的则指团形折扇。团扇有圆形、心形和多边形。

19 世纪中叶，广州最高产的制扇人是浩晟，他的店铺位于靖远街 10 号，现存很多外销扇扇盒内盖处都留着他店铺的名称、经营范围和地址。此外，擅长制作象牙扇的“TYSHING”和擅长制作漆扇的“HOUGUA”以及两者兼营的“LINHING”都是名噪西方的制扇名家。

These export fans made in Canton are quite different from the traditional fans in China. They are colorful, unique in craftsmanship, diversified in materials and shapes, and exuberantly beautiful.

China's export fans can be roughly divided into two types: folding fans and round-shaped fans. However, they can be subdivided into different types according to their shapes, such as fans with or without folding surfaces. The fan surfaces can be further divided into symmetry types and asymmetry types, ordinary types and multi-layered types, while the round-shaped fans belong to the types which are made of bones. In terms of round-shaped fans, there are circular shapes, heart shapes and polygon shapes.

In the middle of the 19th century, the most productive fan maker in Canton was Haosheng. His shop was located at No. 10, Jingyuan Street. The name, business scope and address of his shop can still be found on inner side of the cover of many

November / 2021.11

19

星期五

Friday

辛丑牛年 己亥月

十月十五

下元节

Xia Yuan Festival

export fan boxes. In addition, “TYSHING”, who was good at making Ivory fans, and “HOUGUA”, who was good at making lacquer fans, and “LINHING”, who made fans of both types, were fan makers famous in the west.

November / 2021.11

20

星期六

Saturday

辛丑牛年 己亥月

十月十六

19 世纪象牙柄绣花卉纹缎面执扇。

Folding fan with ivory handle and floral patterns on satin, 19th century.

清代双面广绣贴象牙面人物花鸟图折扇，配黑漆描金人物盒。18支玳瑁扇骨，两大骨剔地浮雕柳亭人物花卉纹；小骨双面透雕拉丝锦地，浅浮雕花卉、庭院人物通景画；米色缎扇面，双面彩色、金色丝线绣贴象牙面垂钓人物、花鸟纹样。绣画设色淡雅，在颜色深沉的扇骨衬托下，华而不俗。其材料珍贵，玳瑁透雕与广绣工艺相结合，雕工、绣工精细。

Folding fan made of hawksbill shell with figural patterns on Canton embroidery, Qing dynasty. The fan, made of hawksbill shell, consists of 18 sticks, with two large sticks carved with low-relief flowers, characters, willows and pavilions; the small sticks are carved on both sides with pierced brocade ground on which there are low-relief flowers and characters in the landscape courtyard; both sides of the beige satin are embroidered with colourful and golden fish, flowers, birds and figures who are fishing. The colours of the embroidery are light and elegant, and look gorgeous but not vulgar in contrast to the deep colour of the fan sticks. The precious materials, such as the hawksbill shells, are combined with the Cantonese embroidery techniques. The carving technique is refined and delicate.

November / 2021.11

21

星期日

Sunday

辛丑牛年 己亥月

十月十七

骨扇开生面

Fans Made of Bones

骨扇指没有扇面，完全由扇骨组成的扇子。骨扇并非中国本土出产的扇子，它最早出现在1000多年前的日本奈良时代，于16世纪初通过全球贸易航路从日本传至欧洲，由意大利工匠在日本桧扇的基础上创造出扇面可以折叠的骨扇，再被西方商人带到广州。骨扇是目前已知最早的清代外销扇，约在17世纪晚期大量销往欧洲，风行200余年。

中国骨扇的材质有象牙、玳瑁、砗磲、檀香木、描金漆和金银累丝几种，其工艺集镂通雕刻、金漆彩绘、劈丝编缀、螺钿镶嵌、银胎烧珐琅等于一体，充分展现了广作工艺的精华。此外还有一种扇面可360度展开的圆形骨扇，但因工艺复杂，价格昂贵，仅在18世纪末至19世纪初短暂生产后就停产了。

A fan made of bone refers to a fan without a mounted surface, and is completely made of bones. Fans made of bones did not firstly appear in China, but in the era of Nara in Japan, more than a thousand years ago. At the beginning of the 16th century, they were introduced to Europe from Japan through the global trade routes. Based on the Japanese wooden fans, the Italian craftsmen created the folding fans made of bones, which were then brought to Canton by Western merchants. Fans made of bones were the earliest known export fans in the Qing dynasty, and in the late 17th century they were sold to Europe in large amount and became popular for more than 200 years.

The materials of the fan sticks made in China can be divided into several types: ivory, hawksbill shells, giant clam, sandalwood, golden lacquer and filigree with gold and silver. Different techniques such as pierced-carving, gold-painting, splitting, weaving, mother-of-pearl inlay and enameling on silver were combined together to

小雪

Slight Snow

November / 2021.11

22

星期一

Monday

辛丑牛年 己亥月

十月十八

show the quintessence of Cantonese craftsmanship. In addition, there is a kind of round-shaped fans made of bones whose stretching angle is up to 360 degree, but due to the complex techniques and high prices, fans of this type were only produced for a short period in the late 18th century and early 19th century and then the production stopped.

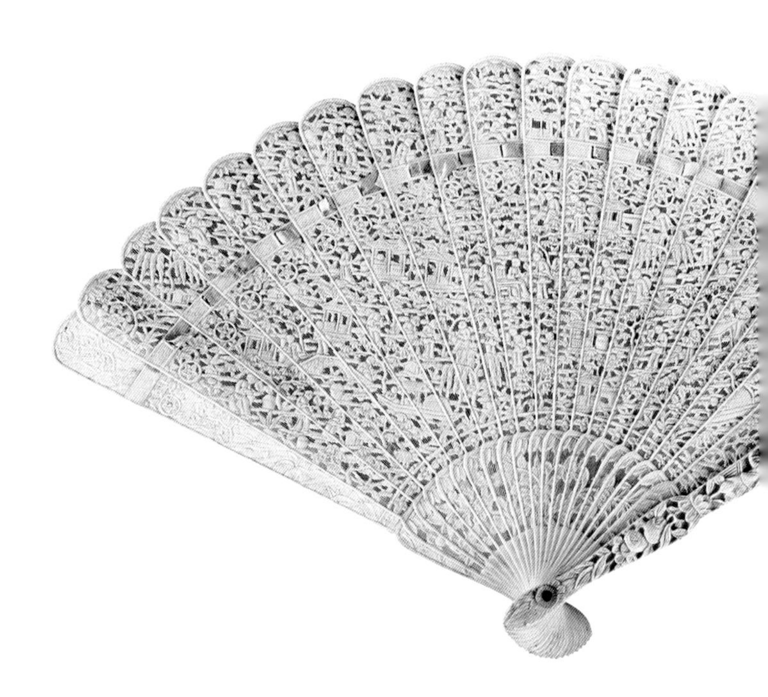

19世纪象牙镂雕扇骨人物纹折扇。（原件藏于广州十三行博物馆）

Ivory folding fan with pierced decorations and figural patterns, 19th century. (The original piece is collected in Guangzhou Thirteen Hongs Museum.)

November / 2021.11

23

星期二

Tuesday

辛丑牛年 己亥月

十月十九

牙扇有仙工

Excellent Craftsmanship of Making Ivory Fans

象牙扇可以说是外销扇中的“霸主”，中国最早出口欧洲的折扇主要材质便是象牙。从 17 世纪晚期起，象牙扇流行了将近百年，造型、材质丰富多样，几乎涵盖了外销扇的所有品种，包括折扇和团扇。在材质方面，既有纯象牙扇，也有象牙为骨，纸面、广绣、羽毛等为扇面的扇子。工艺方面就更多了，有劈丝、编缀、透镂雕、浮雕、彩绘、油画、镶嵌、象牙骨牙片贴面等。这其中，以“牙丝宫扇”最为有名。工匠将象牙以药水软化后，劈成直径不足一毫米的细丝编织成扇。牙丝宫扇是广东独有的地方贡品，有着“南派象牙仙工”的美誉。

Ivory fans can be deemed as “the most delicate” among all the export fans. The earliest fans exported to Europe from China were made of ivory. Since the late 17th century, ivory fans had been popular for nearly a hundred years. Made of various materials with different shapes, the ivory fans can be found in all kinds of export fans, such as folding fans and round-shaped fans. In terms of materials, there are not only Ivory fans, but also fans with ivory sticks and mounted with paper, embroidery, feathers and so on. There are even more types of techniques, such as splitting, weaving, pierced-carving, low-relief, polychrome painting, oil painting, inlaying, attaching ivory and bone veneers, etc. Among them, “fans made for the court by ivory threads” were the most famous. After softening the ivory with liquid medicine, the craftsman split the ivory into thin threads less than one millimeter and weaved them into a fan. “Fans made for the court by ivory threads” were unique local tributes from Guangdong province, and known as “the excellent ivory craftsmanship of the Southern school”.

November / 2021.11

24

星期三

Wednesday

辛丑牛年 己亥月

十月二十

清道光贝雕扇骨纸本彩绘贴象牙面戏剧故事图折扇。（原件藏于广州十三行博物馆）

Folding fan made of shell with polychrome illustration of Chinese traditional opera, reign of Emperor of Daoguang, Qing dynasty. (The original piece is collected in Guangzhou Thirteen Hongs Museum.)

November / 2021.11

25

星期四

Thursday

辛丑牛年 己亥月

十月廿一

檀扇香风醉

Fans Made of Sandalwood

檀香木在中国的粤、桂、琼、滇、闽等地虽有少量分布，但大量的檀香木都来自海外。中国自元代起从东南亚和太平洋诸岛进口檀香木，明清之际在广州形成大规模贸易，用以满足“广作”的原料需求。以檀香木雕刻而成的扇子是 17 至 19 世纪众多中国外销扇中的一种。当时，广州的制扇工匠凭借广州作为檀香木交易市场的优势，率先制作檀香扇。檀香扇的种类有折扇和羽扇，其制作工艺与象牙扇接近，有拉花、透雕、镂通雕和彩绘等。广州檀香扇的外销在民国时期因工艺水平下滑而衰落，被苏杭檀香扇所取代。

Although sandalwood is distributed in Guangdong, Guangxi, Hainan, Yunnan, Fujian and other places in China, a large number of sandalwood is imported from overseas. Since the Yuan dynasty, China had been importing sandalwood from Southeast Asia and the Pacific Islands. During the Ming and Qing dynasties, a large-scale trade was formed in Canton to meet the demand for raw materials to make “Cantonese furniture”. The fans carved from Sandalwood were a type of many Chinese export fans from the 17th to 19th century. At that time, with the advantages of Canton being a sandalwood trading market, the Cantonese craftsmen took the lead in making sandalwood fans. Sandalwood fans were divided into folding fans and feather fans. The production process was similar to that of ivory fans, including making jacquard patterns, hollowed-out carving, pierced-carving, and polychrome painting. During the period of the Republic of China, the export volume of Cantonese sandalwood fans declined due to the sliding craftsmanship, and they were replaced by sandalwood fans made in Suzhou and Hangzhou.

November / 2021.11

26

星期五

Friday

辛丑牛年 己亥月

十月廿二

20世纪初檀香木扇骨透雕花卉纹折扇。

Sandalwood folding fan with pierced patterns of flowers, early 20th century.

November / 2021.11

27

星期六

Saturday

辛丑牛年 己亥月

十月廿三

19世纪骨雕扇骨彩绘花卉人物纹羽毛折扇。

Folding feather fan carved from bone with polychrome floral patterns and figures, 19th century.

November / 2021.11

28

星期日

Sunday

辛丑牛年 己亥月

十月廿四

蓝扇宝光华

Filigree Enamel Fans

在清代出口的众多外销扇中，有一类扇子十分奢华贵重，其观赏性大于实用性，这就是银鎏金累丝烧蓝扇。“银鎏金累丝”和“烧蓝”分别指两种工艺。前者指在银丝上鎏一层黄金，再通过盘曲、掐花、填丝、堆垒等方法制作金银首饰的一种工艺；“烧蓝”即银胎画珐琅，它以银为胎，在胎上用银花丝掐出需要的花纹，再填上透明或半透明的珐琅釉料，经过多次低温焙烧而成。银鎏金累丝烧蓝扇以银鎏金累丝工艺制作扇骨，以烧蓝工艺制作图案，制作过程复杂，成本十分高昂，堪称广州外销扇中的奢侈品，产量并不高，存世量更少，仅在嘉庆年间流行了很短的时间便停止了生产。

Among the various export fans made in the Qing dynasty, there was a kind of fans especially luxurious and valuable, and their ornamental value was greater than practicability. Those were the filigree enamel fans. “Filigree” and “enamel” refer to two crafts respectively. The former refers to a process of gilding a layer of gold on the silver, and then making gold and silver jewelry by winding, pinching, filling, stacking and other methods; “enamel” refers to the silver with painted enamel. The artisans take silver as the objects, pinch out the required patterns on the objects with silver wires, then fill in transparent or translucent enamel glaze, and finally finish the process by firing them at low temperature for several times. The sticks are made of silver with filigree and enamel techniques, while the patterns are enameled. The production process is complex and the cost is very high. So fans of this type can be regarded as a luxury among all the Cantonese export fans. The output of these fans was not high, and the amount of the fans handed down was even less. The production lasted for a short time and stopped during the reign of Emperor Jiaqing.

November / 2021.11

29

星期一

Monday

辛丑牛年 己亥月

十月廿五

清代银鎏金累丝烧珐琅纸本彩绘贴象牙面人物图折扇。16支银鎏金扇骨，两大骨采用银鎏金浅浮雕柳亭人物、石榴、卷叶纹；小骨采用银鎏金勾连纹累丝铺地，饰以烧珐琅花卉纹；纸本扇面，双面彩绘，一面通景彩绘人物官邸家居生活场景，人物面部贴象牙面，立体感强，四周辅以花卉为饰；另一面扇面为同一题材内容。此扇成本昂贵，工艺精湛，风格华丽，集合多种材质和工艺于一身。（原件藏于广州十三行博物馆）

Filigree enamel folding fan with gold-painted polychrome figures, Qing dynasty. The fan is made up of 16 silver gilded sticks. Two of the big sticks are carved with low-relief patterns of willows, pavilions, figures, pomegranates and rolling leaves. The small sticks are covered with gilded filigree patterns and decorated with enamel flowers. There are polychrome paintings on both sides of the paper: one side is painted with a life scene in the official's residence, and the faces of the figures are pasted with ivory, forming a strong three-dimensional sense, while the painting is surrounded by flowers as decorations; the other side is painted with the same theme. This fan is expensive in cost, exquisite in workmanship, and gorgeous in style, integrating a variety of materials and crafts into the same piece. (The original piece is collected in Guangzhou Thirteen Hongs Museum.)

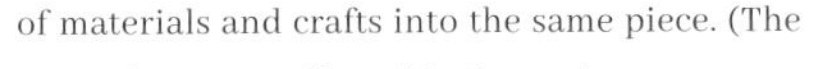

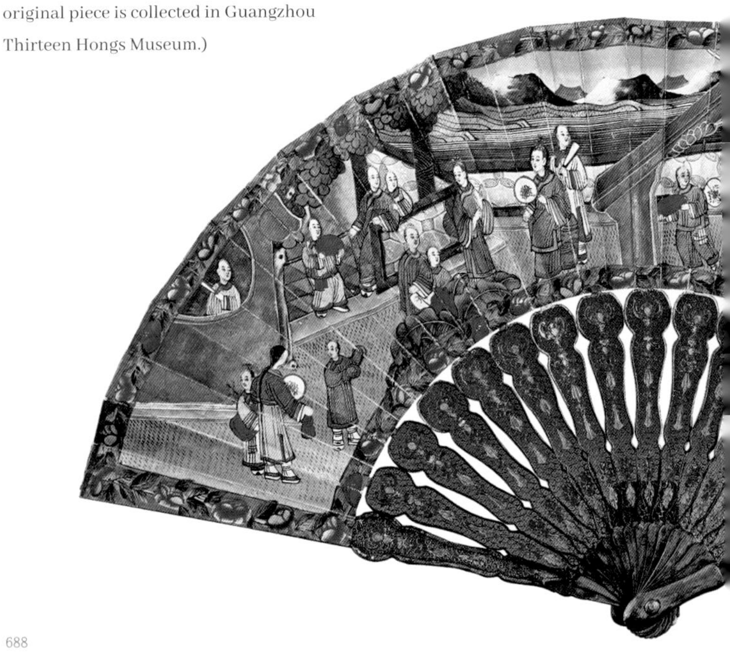

November / 2021.11

30

星期二

Tuesday

辛丑牛年 己亥月

十月廿六

December

一	二	三	四	五	六	日
		1 廿七	2 廿八	3 廿九	4 十一月	5 初二
6 初三	7 大雪	8 初五	9 初六	10 初七	11 初八	12 初九
13 初十	14 十一	15 十二	16 十三	17 十四	18 十五	19 十六
20 十七	21 冬至	22 十九	23 二十	24 廿一	25 廿二	26 廿三
27 廿四	28 廿五	29 廿六	30 廿七	31 廿八		

流星玳瑁扇

Fans Made of Hawksbill Shells

玳瑁是一种生长在热带、亚热带海域的海龟科动物，其龟甲也叫作“玳瑁”，介于半透明到微透明之间，黄色底色上装点绮丽的褐色斑点，色泽鲜丽通透，很早便被中国人当作奇珍异宝用于制作各种首饰、装饰品和乐器零件等。西方人对玳瑁也不陌生，古代埃及、希腊和罗马已经有玳瑁制作的装饰品。

由于玳瑁具有一定韧性，且本身就有光、色和斑纹美，清代广东制扇匠人便将其用作扇子的小骨，以减地平凸等镂刻手法使其更加玲珑剔透，再与象牙大骨相配，使制成的扇子更加富丽雍容。由于材料获取不易和材质较难雕刻，玳瑁扇在清代外销历史上活跃的时间较短，仅存在于1800年到1850年之间，之后便不再生产。

Hawksbills are a kind of turtles that live in the tropical and subtropical sea areas. Their shells are called “hawksbill shells” whose texture is between translucent and slightly transparent, with brown mottling dots scattering on the bright and transparent yellow background. Thus they are seen as rare treasures by Chinese people and were inlaid into all kinds of jewelry, decorations and musical instruments. Westerners are familiar with hawksbill shells. In the ancient Egypt, Greece and Rome, they made ornaments out of hawksbill shells.

Because the hawksbill shells are tenacious, bright, with beautiful colour and patterns, in the Qing dynasty, the Cantonese fan makers used them to make the small fan sticks. They made the fan sticks more exquisite and transparent by using carving techniques such as mezzo-relievo, and then matched them with larger ivory sticks, so that the fans look more colorful and elegant. Due to the difficulty in obtaining materials and the complex carving techniques, fans made of hawksbill

December / 2021.12

1

星期三

Wednesday

辛丑牛年 己亥月

十月廿七

shells, existing between 1800 and 1850, were only popular for a while in the export trade history of the Qing dynasty, and were no longer produced.

玳瑁扇骨绢本孔雀纹折扇。（原件藏于广州十三行博物馆）

Folding fan made of hawksbill shell with patterns of peacocks on silk. (The original piece is collected in Guangzhou Thirteen Hongs Museum.)

December / 2021.12

2

星期四

Thursday

辛丑牛年 己亥月

十月廿八

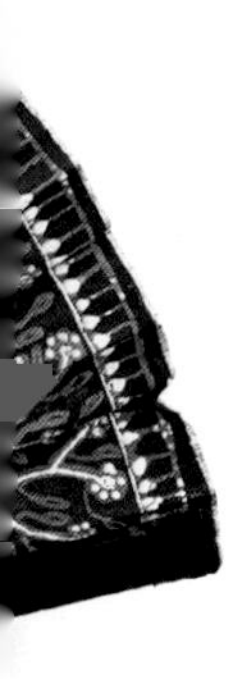

奢华唯漆骨

Lacquer Fans

以木片做扇骨，经髹漆、涂漆胶、扫金粉等多道繁复工序制作而成的扇，叫做漆骨扇。漆骨扇以折扇为主，工艺十分考究。经过螺钿、雕漆、漆上洒金、退光洋漆等工艺制作的漆骨扇，挥扇间流光溢彩，既华丽又生动。漆骨扇大约从 18 世纪末开始销往海外，其工艺以绘画为主。漆骨扇常以黑漆或朱漆为地，以金粉或银粉为颜料绘制纹饰图案。除中国传统的山水、花鸟、人物之外，也有西方来样定制的徽章纹样，在不同时期流行的纹饰不大一样，例如乾嘉之际以葡萄藤蔓纹饰为主，道光早期盛行山水动物纹，道光中晚期多见庭院人物故事题材。其中一种叫做“百脸扇”，以象牙贴片作为人物脸部，再用丝绸剪成人物衣服粘贴于扇面，黑漆描金扇骨，十分华丽。

The lacquer fans made of wood, processed by many complicated steps such as painting, gluing and sprinkling gold powder, are called lacquer bone fans. Lacquer fans are mostly folding fans with exquisite craftsmanship. By inlaying mother of pearl, carving the lacquer, sprinkling gold on the paint, polishing the lacquer, etc., when held in hands, the fans look colourful, brilliant, luxurious and vivid. Lacquer fans had been sold overseas since the end of the 18th century, and the technique applied on them was mainly painting. Lacquer fans were usually painted black or vermilion, on which the patterns would be painted with gold or silver powder. In addition to the traditional Chinese landscapes, flowers and birds, there are also Western customized patterns of coat of arms. The popular patterns in different periods were not the same. For example, during the reign of Emperor Qianlong and Emperor Jiaqing, the main patterns were vines. In the early period of Emperor

Daoguang, the patterns of landscapes and animals prevailed, while in the middle and late period, the characters in the courtyard and stories were commonly taken as the subjects. There is a kind of fans called “the fans with one hundred faces”. The faces of the figures are made of ivory pieces, the clothes of the characters are made of silk and pasted on the surface of the fan, and the fan sticks are lacquered with black colour and painted with golden patterns. The lacquer fans of this type are resplendent.

清代黑漆描金绢本彩绘贴画贴象牙面执扇。

Black lacquer fan with polychrome painting on paper, Qing dynasty.

December / 2021.12

4

星期六

Saturday

辛丑牛年 己亥月

十一月初一

清末黑漆描金庭院人物纹折扇。21 档黑漆描金扇骨，两大骨以金、红两色漆绘福在眼前、八宝等吉祥纹饰；扇面顶部以庭院人物纹为饰；中部主体双凤纹云头形开光内绘庭院休憩图，中间有椭圆形留白徽章；底部饰西洋卷草花卉。此扇采用描金和堆漆两种技法，画面金光灿烂，立体感强。

Black lacquer folding fan with gold-painted figures in the courtyard, late Qing dynasty. There are 21 black lacquer sticks with gold-painted patterns. The two big sticks are painted with gold and red auspicious patterns, such as “happiness before our eyes” and eight precious objects. The top of each fan stick is decorated with figures in courtyards. The main part in the middle is painted with a cloud-shaped open window where there is a painting of figures resting in a courtyard, and which is surrounded by two phoenixes. In the middle of the open window, there is an oval area left for the coat of arms. The lower part of the fan is decorated with the western rolling grasses and flowers. Two techniques, gold-painting and stacking-lacquer, are applied in the making of this fan, whose painting is resplendent with a strong sense of realism.

December / 2021.12

5

星期日

Sunday

辛丑牛年 己亥月

十一月初二

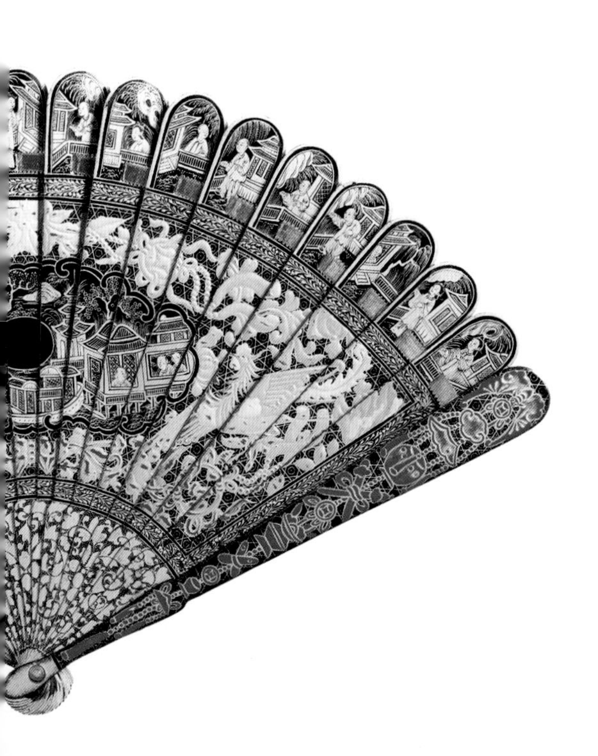

贝扇虽好不坚牢

Fans Carved from Shells

清代的广州手工业发展程度极高，各种令人叹为观止的工艺品层出不穷。例如贝壳不仅被广泛用于螺钿镶嵌中，那些完整的贝壳还被拿来做成贝雕工艺摆件和贝雕扇。工匠们在贝壳上或以通雕、镂刻和浮雕雕出富有立体感的图案，或以细细的线刻绘制出工笔画，再经过打磨、去污、上色、镶嵌、打蜡等工序方才完成一件作品。贝雕扇则除了上述工艺外，还需要将贝壳切割打磨成扇骨，有的还要再结合其他材料组装成扇，工艺更加复杂。贝壳的质地脆薄，且不同品种贝壳特性也有所差异，在制作过程中稍微操作不当就会导致前功尽弃，因此贝雕扇产量极低，留存至今的都代表了当时高超的工艺水平，然而因贝壳制品不易保存，大多有不同程度的损坏。

In the Qing dynasty, the development of handicraft industry in Canton reached a very high level, and a variety of amazing handicrafts emerged one after another. For example, shells were not only widely used as inlaid objects, but also used to make ornaments and fans. Craftsmen carved the realistic patterns on shells by using techniques such as hollowed-carving, pierced-carving and low-relief. They also carved the patterns of meticulous brushwork style with extremely thin lines, and then finished the artwork by polishing, cleaning, painting, inlaying, waxing and other processes. In addition to the processes above, to make the sticks of shell-carving fans, the artisans also needed to have the shell cut and polished, and in some occasion, they had to apply more complicated craft, such as combining other materials to make a whole fan. The shell's texture is brittle and thin, and varies with different categories. A tiny improper operation in the production process might

December / 2021.12

6

星期一

Monday

辛丑牛年 己亥月

十一月初三

result in wasting all the previous achievements. Therefore, the output of the shell-carving fans was very low, and the remaining ones represent the highest technical level of that period. However, the shell products are not easy to preserve, most of them are damaged to some extent.

清末贝壳骨浅浮雕人物风景图折扇。24档贝壳扇骨丝带连缀，两大骨剔地浅浮雕瑞兽、人物、瓶花，小骨采用隐起浅浮雕工艺。一面扇面雕刻渔樵耕读、市井人物，画面紧凑，布局热闹；另一面扇面内饰岭南水乡图，辅以缠枝洋花点缀其间。此扇雕工精细，浓郁的东方风情跃然扇面之上。贝类颜色比对筛选难度较大，因而这类扇产量不高，存世量较少。（原件藏于广州十三行博物馆）

Folding fan made of shell with figures and landscapes, late Qing dynasty. The fan is composed of 24 sticks made of shell and bound by a ribbon. Two big sticks are carved with auspicious animals, figures and flowers in vases; the small sticks are carved with protruding low-relief technique; one side is carved with fishermen, woodcutters, ploughmen and figures in market, with compact arrangement and lively layout; the other side is a Lingnan water town scenery, decorated by entwined branches and flowers. This fan is finely carved and has a strong oriental style. It would be difficult to maintain consistency in texture because each shell has its colour, so the output of this kind of fans was not high, and the number of the survival ones is small. (The original piece is collected in Guangzhou Thirteen Hongs Museum.)

大雪

Great Snow

December / 2021.12

7

星期二

Tuesday

辛丑牛年 庚子月

十一月初四

绣扇见“针”章

Fans Mounted with Embroidery

17 至 18 世纪，欧洲先后流行巴洛克风格和洛可可风格，以轻快、精致、细腻和繁复为特点，崇尚豪华和气派。广州十三行的各种外销品充分迎合西方市场的审美，以色彩绚烂、精致繁复、融汇中西为特点，与在国内销售的同类产品大相径庭。广绣扇是广州制扇工匠应海外市场需求，将广雕和广绣工艺创造性融合起来形成的外销扇品种。从一幅幅外销广绣扇的小小扇面里，可以对那个时代的风尚略窥一二。以清道光透雕象牙骨缎面广绣花鸟纹折扇为例，在米色广缎扇面上，用五彩丝线双面刺绣了花鸟纹，看似淡雅清净的画面囊括了平针、咬针、插针、续针、扭针、渗针、舒针、勒针、洒插针、转文续针等广绣针法，工艺精湛，低调奢华。

In the 17th and 18th century, Baroque and Rococo style, characterized by lightness, delicacy, meticulousness and complexity, successively became popular in Europe, advocating the pursuit for the luxurious and splendid style. All kinds of export products in the trading region of Canton were made to cater to the aesthetic taste of the western market. They were characterized by gorgeous colours, exquisite complexity and integration of Sino-Western elements, and quite different from similar products sold in domestic market. Fans mounted with embroidery were a kind of export fans made to meet the needs of overseas market by Cantonese craftsmen who creatively integrated Cantonese carving and embroidery. From a small fan mounted with embroidery sold abroad, we can have a glimpse of the fashion of that period. The light and transparent ivory folding fan mounted by Cantonese embroidery with patterns of flowers and birds made in the period of Emperor Daoguang in the Qing dynasty can be taken as an example. On the cream-

December / 2021.12

8

星期三

Wednesday

辛丑牛年 庚子月

十一月初五

coloured embroidery mounted on the fan, the flowers and birds are embroidered with multicoloured silk threads on both sides, which look elegant and neat, and the techniques applied to it include threading the Ping needle, Yao needle, Cha needle, Xu needle, Niu needle, Shen needle, Shu needle, Lei needle, Sacha needle, Zhuanwenxu needle and other techniques of Cantonese embroidery. This embroidery fan is made with exquisite workmanship and looks luxurious but not emphatique.

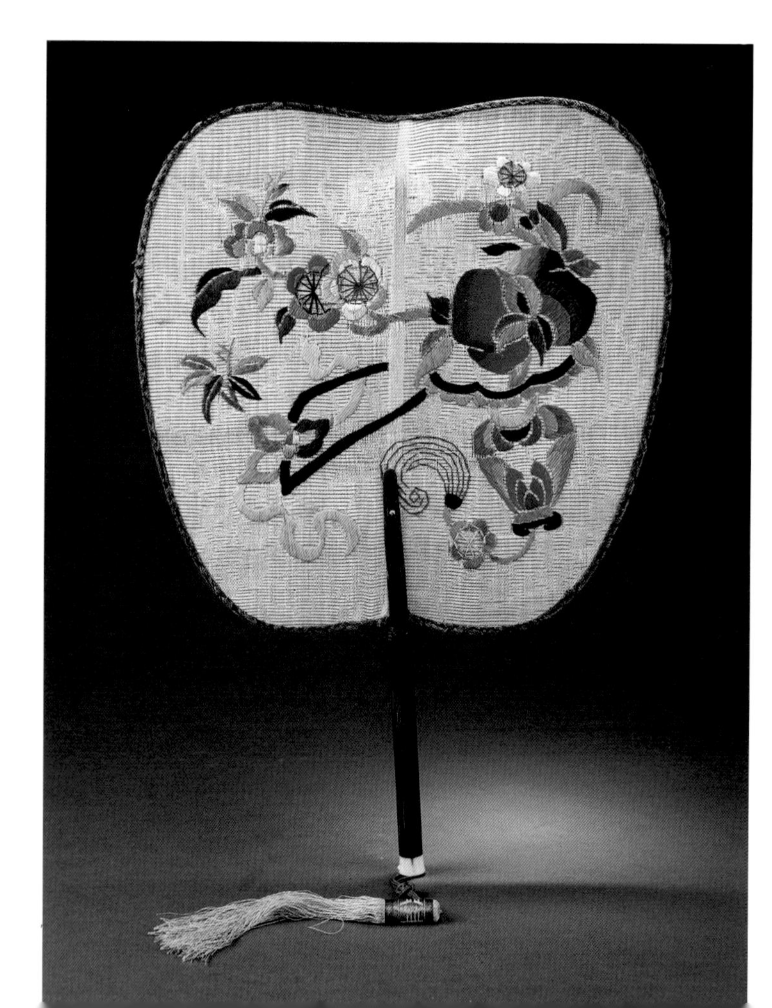

December / 2021.12

9

星期四

Thursday

辛丑牛年 庚子月

十一月初六

清光绪竹柄绣花卉寿桃纹执扇。（原件藏于广州十三行博物馆）

Fan made of bamboo with patterns of flowers and peaches, reign of Emperor of Guangxu, Qing dynasty. (The original piece is collected in Guangzhou Thirteen Hongs Museum.)

扇面广风情

Chinese Aura on the Fans

清代广州外销扇可以说几乎集纳了所有经典的广州工艺，包括雕刻、编织、镶嵌、珐琅、刺绣、髹漆和彩绘等，其中的彩绘扇面，无论在绘制技法还是在题材方面，都与当时的外销画同为一脉。

外销扇的彩绘颜料种类多样，有水粉、水彩、泥金等，绘制技法则在中国画基础上引入西方的明暗对比、透视和色彩晕染，具有强烈的时代和地域特征。在画面布局和底纹装饰方面，彩绘扇面又吸收了广彩的纹饰造型特点，如云纹、卷草纹、折枝花纹、双弦纹、勾连回纹、开光等的运用，使布局丰满而热闹。画面的题材也很广泛，花鸟、山水、人物、神话、传说、建筑、戏曲等无所不及，向西方人展现了一幅幅清代中国的风情小品。

It can be said that Cantonese export fans made in the Qing dynasty embody almost all the classical Cantonese crafts, including carving, weaving, inlaying, enameling, embroidery, lacquering and polychrome painting, among which, the fans with polychrome painting bear similarity with the export paintings at that time in terms of the painting techniques and subjects.

There are various kinds of colourful pigments for painting export fans, such as gouache, watercolour, mud-gold (coating material made of glue and gold powder), etc. Based on the Chinese painting, the artisans applied the techniques introduced from the west, such as chiaroscuro, perspective and sfumato, so that the fans bear a strong temporal and regional characteristic. In terms of layout and decorative based-patterns, the characteristics of the Cantonese porcelain patterns, such as clouds, curling grasses, twigs, two strings, fretted patterns, open windows, etc. were also added to decorate the polychrome surface, making the composition full

December / 2021.12

10

星期五

Friday

辛丑牛年 庚子月

十一月初七

and lively. The subjects of the paintings are all inclusive, such as flowers and birds, landscapes, characters, myths, legends, architectures, operas and so on, providing the westerners with sketches of Chinese aura of the Qing dynasty.

清光绪黑漆描金扇骨纸本彩绘贴象牙人物纹折扇。

Black lacquer folding fan with gold-painted and polychrome figural patterns, reign of Emperor of Guangxu, Qing dynasty.

December / 2021.12

11

星期六

Saturday

辛丑牛年 庚子月

十一月初八

清光绪骨雕扇骨彩绘花卉纹羽毛折扇。

Folding feather fan made of carved bone with polychrome floral patterns, reign of Emperor of Guangxu, Qing dynasty.

December / 2021.12

12

星期日

Sunday

辛丑牛年 庚子月

十一月初九

羽扇轻摇东海岸

Feather Fans

用禽鸟羽毛制作的扇子称为羽扇，是中国最古老的扇子之一，在魏晋南北朝之前作为礼仪用具，南北朝以后出现实用型羽扇，一直流传至今。羽扇通常用鹅毛制成，此外还有雉尾、鹤尾、雕翎、鹰翎、孔雀翎、鹊翅等，以颜色纯白为上品，经采羽、选羽、刷羽、洗羽、理毛、修片、缝片、装柄、整排、饰绒等诸多工序制成。羽扇以柄居中，两边用羽对称，一扇用羽从十几羽到二三十羽，一般以竹签或金属丝穿翎管编排成形，属于不可折叠的平扇。19 世纪中叶，美国是中国外销羽扇的主要市场，当时每把羽扇的价格从 1.2 美元到 13 美元不等。那时美国东海岸的妇女，几乎人手一把广州羽扇才算时尚。

The fans made of birds' feathers are called feather fans. As one of the oldest kind among Chinese fans, they were used as ceremonial props before the Wei, Jin, Nan and Bei dynasties. After that, the practical feather fans appeared and they have been handed down till now. Feather fans are usually made of the feathers of the geese, pheasants' tails, cranes' tails, vultures' wings, eagles' wings, peacocks' wings, magpies' wings, and so on. The purely white feathers are deemed as the top grade. The feathers will be processed with many steps such as plucking, selecting, brushing, cleaning, trimming, patching, sewing, assembling into the fan sticks, arranging, decorating, etc. The handle will be placed in the center of a feather fan, while the feathers are added symmetrically on both sides. A fan consists of a dozen to twenty or thirty feathers, which are usually arranged by bamboo sticks or metal wires through the feather tubes. The feather fans are not foldable. In the middle of the 19th century, Chinese feather fans were mainly sold to the American market. At that time, the price of each feather fan cost from $1.2 to $13. Nearly every woman on the east coast of the United States had a feather fan from Canton.

December / 2021.12

13

星期一

Monday

辛丑牛年 庚子月

十一月初十

19世纪初期的外销水粉画：广州钟表店。（图片来源：FOTOE图片库）

A gouache painting of the early 19th century: a Cantonese clock shop. (The picture is bought from the photo gallery FOTOE.)

December / 2021.12

14

星期二

Tuesday

辛丑牛年 庚子月

十一月十一

西洋钟表“广州造” Ⅰ

Western Clocks Made in Canton Ⅰ

明清之际，以“天朝上国”自居的王朝统治者虽对外国的科学技术不以为意，但对其中的珍奇玩意兴趣颇大。例如，明清皇帝都很喜欢西洋自鸣钟。来自上层的喜好很快促使广州诞生了全新的手工业门类——钟表制作，由广州工匠制作的钟表，称为“广钟”。

广州的钟表行最初只是对舶来的西洋钟进行改造。负责采办贡品的广东地方官员和十三行行商延请欧洲的钟表名匠到广州，按照皇帝的喜好和审美改造西洋钟表，比如将表盘的罗马数字换成汉字，把装饰换成符合中国人审美的图案，以及加镶各种珠宝等。当时有不少广州工匠和西洋名匠一起工作，他们后来成为中国最早的一批钟表匠人。

The rulers of the Ming and Qing dynasties were proud that they lived in “the kingdom of heaven”, and did not take foreign science and technology seriously, but they were very interested in the rare and precious objects. For example, the emperors of the Ming and Qing dynasties all loved the Western chiming clocks. The preferences of the upper class soon led to the birth of a new handicraft industry——clock manufacture in Canton. The clocks made by Cantonese craftsmen are called “Cantonese clocks”.

At first, the clock shops in Canton only re-adjusted the imported western clocks. The local officials of Guangdong province and the Hong merchants in charge of the purchase of tributes invited European clock-makers to Canton to transform the western clocks and watches according to the emperor’s preferences and tastes. For example, they changed the Roman numerals of the dial into Chinese characters,

December / 2021.12

15

星期三

Wednesday

辛丑牛年 庚子月

十一月十二

replaced the decoration with patterns catering to the Chinese aesthetics, and inlaid various kinds of jewelry. At that time, many Cantonese craftsmen worked together with the famous western craftsmen, and later became the first batch of clock-smiths in China.

铜镀金珐琅群仙拜寿楼式钟，清乾隆时期广州制造。（图片来源：FOTOE 图片库）

This is a gilded clock made in Canton in the reign of Emperor of Qianlong, Qing dynasty. (The picture is bought from the photo gallery FOTOE.)

December / 2021.12

16

星期四

Thursday

辛丑牛年 庚子月

十一月十三

西洋钟表“广州造” Ⅱ

Western Clocks Made in Canton Ⅱ

通过长期与西洋钟表匠一起工作、学习，广州工匠渐渐掌握了制造钟表的技术，不再局限于对成品的改造。他们先造出了西洋样式的钟表，后来又结合本土工艺特色制造出新型钟表——广钟。广钟保留了西洋钟走时、打点与奏乐的功能，还添加了些极富中国特色的创意，如转花、水法、跑船、对联展开、群猴献桃祝寿等。据说当时的广州聚集了一批世界顶级的钟表工匠，他们制作的广钟内部机芯结构的复杂程度不输西洋钟。在外观上，广钟将岭南特征体现得淋漓尽致，造型多为中式亭台楼阁，还会加装各种体现吉祥富贵寓意的装饰。广钟主要面向国内市场，基本不外销。当时的广钟全国闻名，连清宫造办处自鸣钟处的工匠都以广州工匠为主。

Through long-term co-working and study with the western clock-smiths, Cantonese craftsmen gradually mastered the techniques of making clocks, and were no longer confined to transforming the finished products. They first made clocks and watches of the western style, and then combined them with the local technical characteristics to create a new type of clocks——Cantonese clocks. Cantonese clocks retained the functions of the western clocks such as telling time, striking and playing music, and were also added with some creative ideas of Chinese characteristics, such as the turning flowers, water sprays, sailing boats, expanding couplets, monkeys offering peaches for birthday celebration, etc. It is said that a group of the world's top clock-smiths gathered in Canton at that time, and the inner structures of the Cantonese clocks made by them were no less complicated than those made in the west. In appearance, Cantonese clocks incisively and vividly embody the Lingnan characteristics, with terraces and pavilions of Chinese style and

December / 2021.12

17

星期五

Friday

辛丑牛年 庚子月

十一月十四

various decorations symbolizing auspiciousness and wealth. Cantonese clocks were mainly made for domestic market, not for export. At that time, Cantonese clocks were famous all over the country and even the craftsmen who made chiming clocks in the royal workshop of the Qing court were mostly from Canton.

December / 2021.12

18

星期六

Saturday

辛丑牛年 庚子月

十一月十五

清同治广彩开光花鸟人物纹葫芦形镶铜座钟。这座葫芦形座钟是由一件用于装饰的广彩瓷改造而成，钟表是后来添置的。上腹开窗绘庭院人物图，瓶肩处对称置两铜鎏金人面耳，下腹开窗绘花鸟蝶纹，正中置一圆形钟表。座钟中西合璧，雍容华贵。

Cucurbit-shaped clock with patterns of flowers, birds and figures, reign of Emperor of Tongzhi, Qing dynasty. This cucurbit-shaped clock is transformed from a Guangcai decorative porcelain. The clock was added later. There are figures in the courtyard painted in the open window on the upper part, and two copper gilded handles of human faces are symmetrically arranged on both sides of the porcelain. Flowers, birds and butterflies are painted in the open window on the lower part, and a round clock is arranged in the middle. With a Sino-Western style, the clock looks luxurious and graceful.

December / 2021.12

19

星期日

Sunday

辛丑牛年 庚子月

十一月十六

18 世纪法国布尔工艺铜镶漆木座钟。座钟为机械两套钟，罗马数字珐琅盘，工匠将原本用于宫廷家具的布尔镶嵌工艺运用在钟壳上。座钟是当时欧洲家具的重要组成部分。

Wooden clock inlaid with copper decorations made in France, 18th century. This is a mechanical clock; there are Roman numbers on the enamel dial. The craftsman applied the Beauvais inlay technique, which was originally used to decorate the furniture made for the royal court, on the clock shell. The clocks with stands were important parts of European furniture at that time.

中国茶改变世界——第一个喝茶的欧洲国家

Chinese Tea Has Changed the World——the First European Country to Drink Tea

茶叶成为大宗外销商品是相对晚近的事。在东西航路打通之前，欧洲人没有接触过茶叶。葡萄牙是最早接触到茶叶的欧洲国家，但最早与中国进行茶叶贸易的是荷兰人。1607 年，第一批茶叶从澳门运至印尼万丹，再于 1610 年运回荷兰，拉开了中欧茶叶贸易的序幕。当时，茶叶的价格十分昂贵，但依然深受荷兰贵族和富裕阶层的欢迎，他们甚至将拥有茶叶视为财富的象征。荷兰人将中国的茶器和日本的茶道一并带到了荷兰，还发明了一种独特的饮茶方式——在茶中放入砂糖和香辛料藏红花，形成了独自一派的茶风。1690 年起，荷兰人在爪哇开辟了茶园，茶叶因此得以进入寻常人家，很快荷兰便成为欧洲第一个全民喝茶的国家。

It was relatively late for tea to become an export commodity. Before the opening of the routes which connected the East with the West, the Europeans had no idea about tea. Portugal was the first country to discover tea, but the Dutch were the first to conduct tea trade. In 1607, the first batch of tea was transported from Macao to Banten, Indonesia, and then back to the Netherlands in 1610, which was the prelude of tea trade between China and Europe. At that time, the price of tea was very high, but it was still extremely popular among the Dutch aristocrats and the rich, who even regarded the possession of tea as a symbol of wealth. The Dutch brought Chinese tea utensils and Japanese tea ceremony back to the Netherlands. They also invented a unique way of drinking tea——adding sugar and saffron into

December / 2021.12

20

星期一

Monday

辛丑牛年 庚子月

十一月十七

tea, and formed a special tea culture. In 1690, the Dutch exploited a tea garden in Java, which enabled tea to go into the ordinary families. Soon, the Netherlands became the first European country where all the citizens drank tea.

冬至

Winter Solstice

December / 2021.12

21

星期二

Tuesday

辛丑牛年 庚子月

十一月十八

这幅 18 世纪的外销画《踩茶图》描绘了外商在广州采购茶叶的情景。（图片来源：FOTOE 图片库）

Stepping on the fresh tea leaves, an export painting of the 18th century. The painting depicts the scene where foreign merchants were buying tea in Canton. (The picture is bought from the photo gallery FOTOE.)

中国茶改变世界——荷兰的“中国委员会”

Chinese Tea Has Changed the World——The “Chinese Committee” in the Netherlands

荷兰人不仅自己喝茶，还积极将茶叶推销到邻近国家。从 17 世纪中期起，荷兰先后把茶叶传到葡萄牙、法国、英国、德国乃至北美地区。18 世纪的欧洲已经饮茶成风，茶叶贸易的巨大利润吸引了英国、法国、丹麦、比利时等国也加入茶叶贸易行列。他们采取直接对华贸易策略，让荷兰商人产生了巨大的危机感。在这之前，中荷茶叶贸易采用中国—爪哇巴达维亚—荷兰的间接贸易形式，茶叶转运速度很慢。在受到茶叶倾销冲击后，18 世纪 20 年代，荷兰东印度公司开辟了对华直接贸易航线，并在后来还专门成立了负责对华贸易事务的“中国委员会”。18 世纪中期，茶叶贸易一度占到中荷贸易的 78.9% 至 89.6%。

The Dutch not only drank tea in its own country, but also actively promoted tea in the neighbouring countries. Since the mid-17th century, the Netherlands had successively introduced tea to Portugal, France, Britain, Germany and even North America. In the 18th century, tea became popular in Europe. The huge profits of tea trade attracted Britain, France, Denmark, Belgium and other countries to take part in it. They directly traded with China and made the Dutch traders feel threatened. Before that, the tea trade was conducted indirectly: the tea would be sent to Batavia on Java Island and then transported to the Netherlands, so the trading process was very slow. After the impact of tea dumping, in the 1820s, the Dutch East India Company opened a direct trade route to China, and later deliberately set up a “Chinese Committee” responsible for the trade affairs with China. In the mid-18th century, tea trade once accounted for 78.9% to 89.6% in the whole Sino-Dutch trade.

December / 2021.12

22

星期三

Wednesday

辛丑牛年 庚子月

十一月十九

绘于18世纪初期的纸本水彩画：十三行茶商向法国人出售茶叶。（图片来源：FOTOE图片库）

A gouache painting of the early 18th century, depicting the scene where a Cantonese Hong merchant was selling tea to a French merchant. (The picture is bought from the photo gallery FOTOE.)

December / 2021.12

23

星期四

Thursday

辛丑牛年 庚子月

十一月二十

中国茶改变世界——英式下午茶和鸦片战争

Chinese Tea Has Changed the World——English Afternoon Tea and the First Opium War

今天的英式下午茶已经成为英国文化的一个象征，但与欧洲其他国家相比，英国对茶的接受过程十分缓慢。最初，英国人把茶当作药物售卖。直到 1662 年，带着一箱茶叶嫁给英王查理二世的葡萄牙公主凯瑟琳·布拉甘扎把饮茶习惯带到英国，在英国宫廷和上流社会形成风尚，后来又传至平民社会，渐渐形成全民喝下午茶的生活习俗。英国东印度公司在 1689 年开启了与中国的茶叶贸易。进入 18 世纪，英国对茶叶的需求量呈现出惊人的增长，到 18 世纪末，英国每年要从中国进口大约 2000 万磅茶叶，大量的白银从英国流入中国。为了扭转贸易逆差，英国人开始向中国走私鸦片，随后引发了第一次鸦片战争。

Today's British afternoon tea has become a symbol of British culture. But compared with other European countries, Britain accepted tea in a very slow process. At first, the British sold tea as medicine. It was not until 1662 that the Portuguese princess Catherine de Braganza brought a chest of tea to Britain when she married King Charles II. Thus the habit of drinking tea was introduced to the British court as well as the upper class and became a fashion. Later on, the habit of drinking tea was spread to the civilian society, and gradually became a daily routine of the people all over the country. The English East India Company began to conduct tea trade with China in 1689. In the 18th century, Britain's demand for tea increased tremendously. By the end of the 18th century, Britain imported about 20 million pounds of tea from China every year, and a large amount of silver flowed into China from Britain. In order to reverse the trade deficit, the British began to smuggle opium into China, which led to the First Opium War.

December / 2021.12

24

星期五

Friday

辛丑牛年 庚子月

十一月廿一

December / 2021.12

25

星期六

Saturday

辛丑牛年 庚子月

十一月廿二

清乾隆广彩人物纹茶叶罐。

Tea can with gold-painted patterns of figures, , reign of Emperor of Qianlong, Qing dynasty.

19世纪末清代外销通草水彩画茶文化图，描绘了从炒茶到卖茶的过程。

Tea culture, watercolour on pith paper, late 19th century. The set of export pith paintings depicts the process of stir-firing and selling tea.

December / 2021.12

26

星期日

Sunday

辛丑牛年 庚子月

十一月廿三

中国茶改变世界——茶叶与美国独立

Chinese Tea Has Changed the World——Tea and the Independence of the United States

与英国人对茶叶的狂热相反，向来奉行自给自足小农经济的中国市场对英国货没有太大兴趣，而且中国实行银本位制度，只接受白银交易。当时全球最大的白银产地在美洲，英国向中国买茶叶支付的白银，绝大多数来自他们的北美殖民地。18 世纪 70 年代，随着美洲白银产量的减少和英国本土对茶叶需求量的扩大，如何获取白银成为英国政府面临的问题。为了获取尽可能多的白银，英国先后在北美地区颁布了《印花税法案》《唐森德税法》《茶税法》及《强制法令》，不但加收对北美殖民地的进口税，还确立了东印度公司在北美茶叶贸易的垄断地位。这一系列法令引发了 1773 年波士顿倾茶事件，成为 1775 年美国独立战争的先声。

Contrary to the British enthusiasm for tea, the Chinese market, which had always been a self-sufficient small-scale peasant economy, had little interest in the British goods. Moreover, the currency system in China was based on silver and the Chinese only accepted silver in transactions. At that time, the world's largest source of silver was in North America. Most of the silver that Britain paid for the Chinese tea came from their North American colonies. In the 1870s, with the decrease of silver production in America and the increase of tea demand in Britain, how to obtain silver became a problem faced by the British government. In order to obtain as much silver as possible, the UK successively promulgated *the Stamp Act*, *the Townshend Acts, the Tea Act and the Lutolerable Acts* in North America which not

December / 2021.12

27

星期一

Monday

辛丑牛年 庚子月

十一月廿四

only increased the import tax on the North American colonies, but also ensured the monopoly of the English East India Company in the North American tea trade. These acts triggered the "Destruction of Tea in Boston" in 1773 and became the foreshadow of the American Revolution in 1775.

19世纪通草水彩茶叶贸易图。（原件藏于广州十三行博物馆）

Tea trade, watercolour on pith paper, 19th century. (The original piece is collected in Guangzhou Thirteen Hongs Museum.)

December / 2021.12

28

星期二

Tuesday

辛丑牛年 庚子月

十一月廿五

开放融通铸城魂

Tea Spirit of Open-Mindedness

十三行对近现代广州乃至环珠江口城市文化和气质的形成有着深远的影响，囊括了政治、经济、文化、教育、艺术、城市功能以及社会生活等方面。今天，当我们用一系列词汇来形容岭南地区的文化特征时，会发现它们大多数来自于十三行时期。广东地区负山带海、中有江河勾连的自然地理格局，不仅赋予其浓厚的海洋文化基因，更注定了它必然成为历代王朝观察外界和与外界交流的窗口之一。十三行兴起于特殊的历史政治背景下，整个国家所有对外贸易都集中于一个城市，导致了财富和人才的集中，继而改变了城市的结构和功能。这种机遇此前未见，此后亦不会再有，这也解释了在广东地区长达两千年的对外贸易史中，为何唯有十三行时期在岭南地区的所有方面留下了深刻的印记。

The Canton trade has a profound influence on the formation of the urban culture and atmosphere in modern Canton and even around the Pearl River estuary area in terms of politics, economy, culture, education, art, urban functions and social life. Today, when we adopt a series of words to describe the cultural characteristics of the Lingnan region we will find that most of them had been formed in the Canton trade period. The natural geography of Guangdong province, with the ocean in the front, mountains in the back, and rivers connecting with one another in the middle, not only bears strong genes of maritime culture, but is also bound to establish itself as one of the windows for observing and communicating with the outside world throughout successive dynasties. With such a special historical and political background, the whole country's foreign trade gathered in this city, so that the rich and talented were all gathered here, which had changed the structure and function of the city. This kind of opportunities had never been seen before and will never

December / 2021.12

29

星期三

Wednesday

辛丑牛年 庚子月

十一月廿六

appear again, which also explains why during the two thousand year history of foreign trade in Guangdong province, only the Canton trade period has left such a influential record on all aspects of the Lingnan region.

店铺林立的广州十三行。（图片采自《昔日乡情》）

There were many shops in the trading region in Canton. (The picture is taken from *The Nostalgia for the Hometown*.)

December / 2021.12

30

星期四

Thursday

辛丑牛年 庚子月

十一月廿七

兼容并蓄造岭南

The Inclusiveness of Lingnan Culture

在岭南文化数千年漫长的发展过程中，十三行赋予了它最为鲜明的时代和地域特征，尤其是艺术和绘画方面。如岭南画派的形成正是受了清代十三行外销画的影响，其最大的特点是在中国画的基础上融合东洋、西洋画法，自创一格。广绣虽源自唐代，但其形象传神，颜色富丽，宛如油画的质感是在十三行时期吸收西洋油画中的透视技法和光线折射原理而形成的……十三行为中国工匠和艺术家们提供了很多来自于世界各地的艺术文化资讯，不管是出于商业贸易的需要，还是单纯地为了追求技艺的精进，都启发了岭南艺术的融合与创新。如今，十三行的历史背影虽已远去，但“一带一路”及粤港澳大湾区的崛起将推动岭南城市群进一步融入全球化体系中。未来，在全球文化的交融与碰撞之中，岭南艺术必将绽放更绚丽的花朵。

In the long development of Lingnan culture that lasts for thousands of years, the Canton trade has endowed it with the most distinctive temporal and regional characteristics, especially in terms of art and paintings. For example, the formation of Lingnan School Painting was precisely influenced by the export paintings of the Canton trade in the Qing dynasty. The most conspicuous feature is to combine the eastern painting methods with the western painting methods on the basis of Chinese painting and create a unique style. Although Cantonese embroidery originated in the Tang dynasty, the vivid images, brilliant colours, and the textures similar to the oil paintings were formed in the Canton trade period by integrating the perspective technique and light refraction principle of the Western oil paintings. The Thirteen Hongs had provided the Chinese craftsmen and artists with a lot of artistic and cultural information from all over the world and had surely inspired the

December / 2021.12

31

星期五

Friday

辛丑牛年 庚子月

十一月廿八

integration and innovation of Lingnan art, whether for the demand of commercial trade or simply for the pursuit of the technical advancement. Now, the Canton trade has already faded into history. But “the Belt and Road Initiative” and the rise of the Greater Bay Area of Guangdong, Hong Kong and Macao will further promote the integration of the Lingnan urban agglomeration into the global system. In the future, thanks to the integration and collision of global culture, Lingnan arts will thrive like blooming flowers.

艺述大湾区

Stories Of Arts In The Greater Bay Area

广州十三行故事

The Thirteen Hongs In Canton

（2021）

IMENT

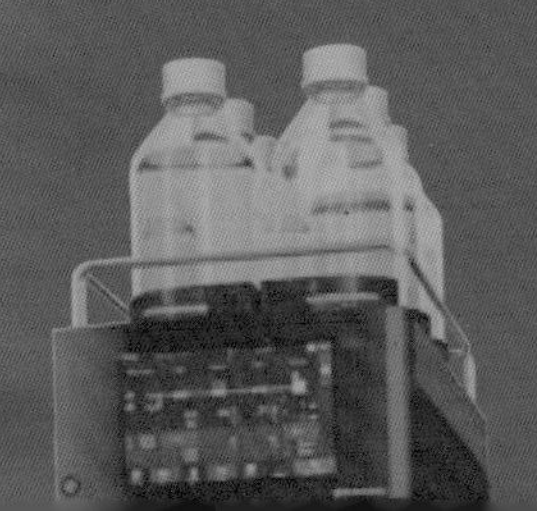